Securing 5G

A Way Forward in the U.S. and China Security Competition

DANIEL GONZALES, JULIA BRACKUP, SPENCER PFEIFER,
TIMOTHY M. BONDS

For more information on this publication, visit **www.rand.org/t/RRA435-4**.

About RAND

The RAND Corporation is a research organization that develops solutions to public policy challenges to help make communities throughout the world safer and more secure, healthier and more prosperous. RAND is nonprofit, nonpartisan, and committed to the public interest. To learn more about RAND, visit www.rand.org.

Research Integrity

Our mission to help improve policy and decisionmaking through research and analysis is enabled through our core values of quality and objectivity and our unwavering commitment to the highest level of integrity and ethical behavior. To help ensure our research and analysis are rigorous, objective, and nonpartisan, we subject our research publications to a robust and exacting quality-assurance process; avoid both the appearance and reality of financial and other conflicts of interest through staff training, project screening, and a policy of mandatory disclosure; and pursue transparency in our research engagements through our commitment to the open publication of our research findings and recommendations, disclosure of the source of funding of published research, and policies to ensure intellectual independence. For more information, visit www.rand.org/about/principles.

RAND's publications do not necessarily reflect the opinions of its research clients and sponsors.

Library of Congress Cataloging-in-Publication Data is available for this publication.
ISBN: 978-1-9774-0855-6

Cover: Tierney/AdobeStock.

About This Report

Fifth-generation (5G) networks are being deployed in the United States and globally and one day will replace many older third- and fourth-generation cellular networks. 5G will provide much higher data rates and lower message latency than any previous generation of cellular network technology. With its advanced capabilities, 5G could enable important new applications. A few possible examples are autonomous connected cars, telemedicine, and augmented reality games and environments. However, security concerns have been raised about 5G networks built using Chinese equipment and 5G phones made by Chinese companies. The United States relies on foreign suppliers for 5G infrastructure and key microchips that go into every 5G phone.

This report examines 5G security issues, the 5G supply chain, and the competitive landscape in 5G equipment and mobile device markets. It describes where U.S. or Chinese companies have technology or market advantages in the emerging 5G security competition between the United States and China. The report provides recommendations for securing U.S. 5G networks and mobile devices and those used by U.S. allies and foreign partner nations.

Funding

Funding for this research was made possible by the independent research and development provisions of RAND's contracts for the operation of its U.S. Department of Defense federally funded research and development centers.

Acknowledgments

Additional support was provided by the RAND National Security Research Division. We have benefited from conversations with other members of the RAND 5G research team: James Bonomo, C. Richard Neu, Samuel Absher, Edward Parker, Jennifer Brookes, Jordan Willcox, David R. Frelinger, and Anita Szafran.

We are also grateful to the reviewers of this report: Edward Balkovich, Jennie W. Wenger, Carter C. Price, and Caitlin Lee of RAND and Marjory S. Blumenthal, senior fellow at the Carnegie Endowment for International Peace. Their thoughtful and thorough reviews have improved this report immeasurably. However, all errors the reader might find are due only to us.

Summary

The Fifth-Generation Security Competition

The competitive landscape in U.S. telecommunications has traditionally been viewed through the lenses of economics and technology, but security issues have become a third major concern. Fifth-generation (5G) networks rely on state-of-the-art microchips. Today, more than 80 percent of microchips made come from Asia, and a growing number are made in China. According to industry projections, by 2030, the United States will make less than 10 percent of the world's microchips, while China and Taiwan together will make more than 40 percent of them (Ip, 2021). Microchips make their way into 5G products in a complex supply chain with many steps and multiple opportunities for microchips to be compromised along the way. For this and other reasons, cybersecurity risks could increase significantly when 5G is deployed worldwide.

China has used Huawei as a means to surreptitiously collect sensitive national security, foreign policy, and intellectual property (IP) information around the world. Huawei was caught spying on the African Union, convicted of stealing software code from U.S. companies, indicted by the U.S. Department of Justice for the theft of U.S. company trade secrets, and assessed to be capable of gathering user data from mobile phones at scale using its equipment deployed in the Netherlands and Belgium. Huawei is subsidized by the Chinese government in many ways and can sell its 5G products at deep discounts that Western competitors cannot match.[1] Huawei and ZTE Corporation, another Chinese 5G supplier, could increase their penetration of the cellular network infrastructure market as 5G is deployed globally, providing China with significant intelligence advantages.

Competition between the United States and China in 5G involves more than economic and technology leadership. Also at stake are the cybersecurity and integrity of next-generation cellular networks. These three dimensions of the 5G competition—economics, technology, and security—are intertwined. The companies that lead in 5G technologies could gain market share, regardless of their security strengths or weaknesses or their trustworthiness, unless cybersecurity and trustworthiness are determined to outweigh technology or economic preferences. An untrustworthy company, such as Huawei, can leverage its market position and government subsidies to help China achieve its national security goals and compromise the security of other nations' networks.

Table S.1 summarizes our analysis of the market, technology, and security positions of Chinese and U.S. companies in 5G. The table shows where Chinese and U.S. companies have a market or technology advantage (designated by A_M or A_T, respectively) or market or technology disadvantages (designated by D_M or D_T, respectively). The 5G architecture areas considered are network infrastructure, mobile devices, mobile device operating systems (OSs), microchip design, and microchip manufacturers (foundries).

As shown in the table, the United States currently has a weak position in two 5G sectors: network infrastructure and microchip foundries—the very top and bottom of the 5G technology stack. In between, in other 5G sectors, the United States has a relative advantage over or is at parity with China. In the rest of this section, we elaborate on these assessments.

5G Network Infrastructure

China has a global market advantage in the 5G network infrastructure market. The United States relies on foreign suppliers—European or Asian companies for 5G infrastructure. In contrast, Huawei and ZTE have

---

[1] By *Western*, we mean democratic and technologically modern.

TABLE S.1

Status of the 5G Security Competition Between China and the United States

5G Architecture	China	United States
Network infrastructure	A_M	D_M
Mobile devices	A_M	A_T
Mobile device OSs	D_T	A_M, A_T
Microchip design	D_T	A_M, A_T
Microchip foundries	D_T	D_T

NOTE: Dark green indicates that the country has a strong advantage in that area. Light green indicates that the country has a slight advantage in that area. Yellow indicates that the country relies on foreign third-party suppliers in that area because it lags in that area but has access to products made by foreign suppliers.

large shares of the global market. There is ample evidence that Huawei's access to state subsidies has given it an unfair competitive advantage. Huawei poses security risks to U.S. allies, developing countries, and rural America because Huawei equipment is still present in some U.S. wireless carrier networks.[2] Huawei and ZTE pose market risks to the two largest suppliers of U.S. 5G infrastructure, Ericsson and Nokia. Both are vulnerable to Chinese competition because they rely entirely on the infrastructure market for profits because both have retreated from the mobile phone market.

Mobile Devices

Chinese and U.S. companies are major players in the mobile phone or device market. Apple is one the largest and most-profitable companies in the world because of the success of the iPhone. In the highly competitive 5G mobile phone market, Apple and Chinese mobile phone makers offer 5G-capable mobile phones. Despite the fact that the United States has blacklisted Huawei, other Chinese phone makers have a greater share of the global market than phone makers from any other country.

Mobile Device Operating Systems

Apple and Google make the two leading mobile device OSs. Most Chinese mobile device manufacturers use Google Android, which enables the United States to use Android as a lever to influence the behavior of Chinese companies. Huawei has been placed on the U.S. Department of Commerce (DOC) entity list. Consequently, Huawei's access to the secure version of Android and the Google Play app store is prohibited, severely impeding its ability to produce 5G mobile phones that can run popular apps. Huawei introduced its own OS in 2020, but, so far, it has limited market share outside China. Other Chinese mobile phone makers have not been blacklisted by the United States and continue to enjoy access to Android. In this 5G market segment, the United States has market and technology advantages.

---

[2] The U.S. infrastructure bill will reportedly include funding for the Federal Communications Commission (FCC) to carry out a "rip-and-replace" program to remove high-risk equipment from the networks of U.S. telecommunication companies.

Microchip Design

U.S. chip designers are market leaders in the microchips needed for 5G networks. Qualcomm is one of the key providers of 5G modem chips. ZTE uses its chips in its base stations, and Chinese 5G phone makers use Qualcomm chipsets. In contrast, Huawei has created its own microchip design company, HiSilicon, that designs many of the chips it needs for its 5G products. Huawei might be catching up to U.S. chip designers, but it still requires Taiwan Semiconductor Manufacturing Company (TSMC) to makes its chips. The DOC blacklisting of Huawei has prevented it from producing its chips at TSMC.

Microchip Foundries

State-of-the-art microchips are needed to meet demanding 5G performance requirements. Microchip foundries that produce microchips with features that are 7 nanometers (nm) or smaller are key links in the 5G supply chain. U.S. and European 5G companies rely on Asian foundries to make 5G microchips. Significant consolidation has occurred in the microchip foundry market as many players, including Intel, have encountered difficulties in manufacturing 5G chips. Only two foundries are presently capable of producing the high-speed logic chips needed for 5G: TSMC and Samsung. Until recently, TSMC supplied chips for both U.S. and Chinese companies. The U.S. blacklisting of Huawei made it illegal for foreign companies to produce chips for Huawei using U.S. technology. U.S. companies provide key chip fabrication equipment to TSMC, so TSMC has complied with the U.S. blacklist order. Samsung and TSMC are currently aligned with U.S. interests. However, if U.S. companies were to lose their market-leading positions in semiconductor manufacturing equipment, there is no guarantee that Samsung and TSMC will comply with U.S. blacklist orders in the future. Even though U.S. companies are no longer competitors in the 5G logic microchip foundry sector, the United States has been able exert pressure on foreign foundries supplying 5G chips to Huawei to create a tenuous advantage in this sector.

Huawei and the Impact of the U.S. Blacklisting

Table S.2 resembles Table S.1 but compares the positions of Huawei and the U.S. ecosystem of 5G companies. Huawei is color-coded green in the infrastructure sector because it supplies 5G networks to many foreign

TABLE S.2

Huawei and U.S. 5G Market and Technology Status

5G Architecture	Huawei	United States
Network infrastructure	A_M	D_M
Mobile devices	D_M	A_T
Mobile device OS	D_M, D_T	A_M, A_T
Microchip design	D_T	A_M, A_T
Microchip foundry	D_M, D_T	D_T

NOTE: Dark green indicates that the entity has an advantage in both market and technology. Light green indicates that the entity has an advantage in either market or technology. Yellow indicates that the entity relies on foreign third-party suppliers but has access to products from these suppliers. Light red indicates that an entity has a disadvantage in either market or technology. Dark red indicates that an entity has a disadvantage in both market and technology.

countries, which gives China a security advantage in those countries, enabling China to collect national security, IP, and foreign policy secrets.

The United States is labeled green in 5G mobile devices for different reasons: The use of U.S.-designed 5G phones reduces the Chinese government's surveillance threat. Huawei is colored light red in this sector because the U.S. blacklisting has prevented it from selling 5G phones in most of the global market.

We assumed that U.S. foreign suppliers were trustworthy and would not intentionally place cyber exploits or back doors into their products, making them less vulnerable to compromise. As before, the United States is color-coded yellow in the 5G network infrastructure and in microchip foundry sectors. Huawei is color-coded red or dark red in all sectors except infrastructure. Because of the U.S. blacklisting, it introduced its own mobile device OS into a market dominated by Apple and Google. Also because of the blacklisting, it has lost access to new U.S. chip design tools and the 5- and 7-nm foundry sectors.

Given Huawei's loss of market share in the mobile phone market, it would not be surprising if the Chinese government has already provided substantial additional financial assistance to Huawei to keep it afloat. Huawei was forced to sell its smartphone division, Honor, to a local government and a Chinese phone distributor, Digital China, for $15 billion, in an all-cash deal in late 2020 (Zhu, 2020). The effects of the U.S. blacklisting have provided a possibly short-term security advantage to the United States relative to Huawei in that it has slowed the deployment of Huawei mobile devices and network infrastructure globally.

If Huawei's mobile OS is adopted by consumers in place of Android, the U.S. blacklisting effort might succeed only until then and could threaten the advertising business model of Google, one of the largest and most-profitable U.S. high-technology companies. It would be especially worrisome if other Chinese mobile phone manufacturers decided to drop Android for the new Huawei OS. To date, there is no evidence that such a move is in the offing, although the Chinese government could direct such a move inside China. Although an effective measure in the short term, keeping the Huawei blacklisting applied to Google Android could be risky in the long term because it could enable Huawei to eventually build an OS platform that is competitive with Android.

The United States has used the chip foundry sector to weaken Huawei's position in the 5G phone market. However, the U.S. security advantage could erode quickly if China is someday able to build state-of-the-art microchip foundries inside China. China has probably already attempted to gain access to TSMC's technology secrets. If it can acquire this technology through cyberattack or espionage, it might be able to establish domestic chip foundries capable of providing Huawei state-of-the-art chips and break Huawei's dependence on TSMC or Samsung. However, this is considered unlikely by most microchip experts, who estimate that China is a decade behind the market leaders (Kessler, 2020). In addition, without state-of-the-art semiconductor fabrication equipment from U.S. and European vendors, it will be even harder for China to build a domestic foundry capable of producing 5G chips.

Recommendations

A U.S. security strategy should secure the integrity of the U.S. 5G microchip supply chain and ensure the technology leadership of trusted network infrastructure providers, without putting at risk the technology leadership of U.S. companies in key 5G market sectors, such as chip design, mobile device OSs, and mobile devices. We recommend that, to achieve these objectives, the United States undertake the activities shown in Table S.3.

TABLE S.3

Current and Recommended 5G-Related Activities for Federal Entities

Organization	Function	Current 5G Activity	Recommended 5G Activity
BIS (part of DOC)	Ensures the U.S. export control and treaty compliance system in support of U.S. strategic tech leadership	Blacklist untrustworthy foreign companies. For example, BIS placed Huawei on the entity list in 2019.	Continue sanctions on Huawei; permit U.S. companies to sell chips to nonsanctioned Chinese phone makers; provide incentives for advanced microchip foundries in the United States.
CISA (part of the U.S. Department of Homeland Security)	Manages cyber and physical risk to U.S. critical infrastructure	Carry out the five lines of effort specified in the CISA 5G strategy.	Monitor 5G supply chains; encourage trusted foreign vendors to adopt CISA cybersecurity SCRM best practices.
DARPA	Conducts basic and applied research	Discover and examine 5G network security and internet-of-things security issues.	Sponsor research on advanced semiconductor fabrication methods and tools; assist DOC in evaluating progress of U.S. companies building 7-nm process foundries in the United States.
FCC	Regulates radio (e.g., cellular), television, wire, satellite, and cable communications	Allocate and auction midband (C-band) spectrum for 5G networks.	License additional midband spectrum; license shared spectrum bands for 5G.
NIST (part of DOC)	Advances science and engineering measurement standards and technology	Maintain the NIST alliance for 5G networks; discover and examine 5G cyber vulnerabilities; write and maintain 5G spectrum-sharing standards.	Continue and expand cybersecurity evaluations of 5G technical standards.
NSF	Conducts basic research	There are no current activities.	Sponsor research on 5G and other advanced wireless communication technologies and on advanced semiconductor fabrication methods and tools.
NTIA (part of DOC)	Advises the President on telecommunication and information policy issues	Write and maintain the 5G implementation plan for *National Strategy to Secure 5G*; conduct research on advanced civilian communication systems.	Increase budget for the Institute for Telecommunication Sciences; with the assistance of DoD and appropriate FFRDCs, develop and evaluate 5G technical standards for spectrum sharing and network security.
USPTO	Grants U.S. patents; registers trademarks; advises the President, the Secretary of Commerce, and federal agencies on IP policy, protection, and enforcement	Review patent applications; conduct patent quality studies.	Link the USPTO patent database with the 5G 3GPP or ETSI technical standard database; evaluate the quality of Chinese 5G patents; examine the links between U.S. and Chinese 5G SEP applications and application timing; provide alerts to the U.S. government if Chinese company–declared SEPs are approved by 3GPP; propose alternative patent application disclosure rules to Congress and the WTO.

SOURCES: BIS, undated; CISA, 2020; DARPA, undated; FCC, undated; NIST, 2021; NSF, undated; NTIA, 2021; USPTO, undated.

NOTE: BIS = Bureau of Industry and Security. CISA = Cybersecurity and Infrastructure Security Agency, part of the U.S. Department of Homeland Security. SCRM = supply chain risk management. DARPA = Defense Advanced Research Projects Agency. NIST = National Institute of Standards and Technology. NSF = National Science Foundation. NTIA = National Telecommunications and Information Administration, part of DOC. DoD = U.S. Department of Defense. FFRDC = federally funded research and development (R&D) center. USPTO = U.S. Patent and Trademark Office. 3GPP = 3rd Generation Partnership Project. ETSI = European Telecommunications Standards Institute. SEP = standard-essential patent. WTO = World Trade Organization.

Assist Trusted Foreign 5G Vendors in Securing Their Supply Chains

The United States should assist Ericsson, Nokia, and Samsung to ensure the security and integrity of their 5G infrastructure products. The U.S. information and communication technology (ICT) industry and CISA have been developing SCRM best practices for securing the supply chain of ICT products. CISA should invite these 5G companies to become members of its ICT SCRM Task Force and work with them to improve 5G SCRM. In addition, the U.S. government should monitor the financial health of these two critical U.S. 5G suppliers.

Allocate More Midband Spectrum to 5G, and Investigate Spectrum Sharing

The FCC should allocate more spectrum at midband frequencies to 5G. Spectrum sharing might be feasible in these bands. DoD and the Institute for Telecommunication Sciences are developing spectrum-sharing technologies and standards. 3GPP is working on spectrum-sharing standards for 5G. A coordinated effort to develop 5G spectrum-sharing standards for the U.S. midband spectrum would be beneficial to the United States and to U.S. companies in the 5G supply chain.

Constrain Huawei's Access to Advanced 5G Chips

Huawei can no longer build 5G phones and reportedly has sold its mobile phone division (although, given its opaque finances, this might, in reality, just be a government bailout). By keeping Huawei on the DOC entity list, the United States will slowly hamper Chinese intelligence collection and IP theft operations around the world.

Exclude U.S. 5G Chip Suppliers from Sanctions

BIS should carefully consider the companies to which U.S. chip designers can sell their chips. Qualcomm, a U.S. company, holds an important technology leadership role in the 5G architecture. It is in the United States' interest to allow Qualcomm to still sell chips to Chinese phone makers that do not provide infrastructure and have not demonstrated a history of espionage. We recommend that Qualcomm still be permitted to sell chips to Chinese phone makers, but the United States must ensure that those chips do not leak into Huawei's supply chain. It might be possible to track Qualcomm chips in the supply chain of Chinese companies if trusted platform module circuitry is embedded in these chips. These chips could be programmed to report their device characteristics when they are first activated. This might make it possible to detect U.S. chips that are activated in Huawei devices.

Permitting Qualcomm, Skyworks Solutions, Qorvo, and other U.S. chipmakers to sell chips to Chinese phone makers helps ensure the economic health of the U.S. 5G supply chain and reduces the economic incentives for the Chinese government to support the development of advanced chipmaking capabilities in China. It would dissociate current U.S. policy toward Huawei from larger, unresolved trade issues between the United States and China.

Protect Technology in U.S. Patent Applications, and Examine Chinese 5G Patents

U.S. policymakers require a better understanding of where and how U.S. and Chinese technologies are incorporated into 5G standards. We recommend that, to do this, the USPTO link its patent database with the 5G 3GPP or ETSI technical standard database. This would enable patent experts to evaluate the quality of Chinese 5G patents submitted as 5G SEPs, identify competing U.S. patent claims, examine the timing of SEP applications, and identify specific areas of 5G technology competition between China and the United States.

The USPTO should also monitor the growing competition between U.S. and Chinese firms in SEP submissions and 3GPP adjudications of submitted SEP declarations. The USPTO should identify overlapping 5G SEPs declared by Chinese and Western firms, the dates such patents are filed, where they are filed, SEP quality, and SEP approval to provide U.S. policymakers indications of whether there is a shift in intellectual leadership in communication technologies from Western to Chinese companies. This information can guide U.S. government–sponsored R&D efforts, to ensure that U.S. companies retain technical leadership in critical areas of wireless communication.

Some experts contend that Chinese companies review U.S. patent applications as soon as they are put online and use these data to file competing patents in China and other countries. This enables China to reduce competition in its domestic market for 5G systems, could slow the approval of declared 5G SEPs filed by U.S. companies, and threatens the integrity of the global patent system. The USPTO should ask Congress to keep patent applications secret until they are approved, like they were prior to 1994. This change would also require a change to WTO regulations, so the administration should immediately raise this issue with the WTO, and push for a commensurate change in WTO regulations governing IP rights and disclosure requirements.

Establish a U.S. Microchip Fabrication Research and Development Program

Both U.S. and Chinese companies remain dependent on two suppliers for advanced 5G chips: TSMC and Samsung. Importantly, the United States and Advanced Semiconductor Materials Lithography in the Netherlands provide the most-advanced tools used in the manufacturing of advanced microchips (Kharpal, 2019). NSF should sponsor research on advanced microchip manufacturing technologies. The United States, with the assistance of its allies, should retain leadership in these areas to ensure control and access to the 5G supply chain. NSF should also fund research in extreme-ultraviolet microchip lithography. This could enable U.S. chipmakers to catch up to Asian industry leaders.

Establish State-of-the-Art Microchip Foundries in the United States

U.S. efforts to encourage TSMC and Samsung to build foundries in the United States should be continued (Davis, O'Keeffe, and Fitch, 2020). At least one foreign-owned foundry should be built in the United States because this will ensure having a trusted source of advanced microchips in case U.S. chipmakers falter.

The United States has only two remaining microchip foundry companies that are near the state of the art: Intel and GlobalFoundries. Both need help. It would be prudent, if they ask, for the U.S. government to provide financial incentives to both to build new state-of-the-art foundries in the United States. These incentives would include those provided by DOC in the Creating Helpful Incentives to Produce Semiconductors (CHIPS) for America Act (U.S. House of Representatives, 2020; U.S. Senate, 2020).

Develop Requirements for Huawei to Enable Its Return to 5G Markets

We recommend that BIS develop a set of conditions for Huawei to meet to demonstrate that it can become a trustworthy partner for 5G. Huawei meeting these conditions would enable U.S. chip designers and foreign foundries to sell their products and services to Huawei. It would also defuse tensions between the United States and China. These conditions should explicitly prohibit IP theft, espionage, and violation of U.S. sanctions.

Areas of Potential Future Research

In this report, we identify areas in which further research is needed to help secure U.S. 5G infrastructure and realize the goals of the U.S. 5G security strategy. The U.S. government should track 5G technology developments and determine which have significant implications for advanced communications. This is highly technical work that several U.S. FFRDCs could pursue.

An important question is whether a government agency or a private organization should be put in charge of a new microchip fabrication R&D program and whether and how costs should be shared with industry. Decades ago, the United States established a private organization called Semiconductor Manufacturing Technology (SEMATECH) to do largely the same thing, initially partly funded by DoD. However, government funding was eventually eliminated. SEMATECH evolved into a private, internationally funded R&D association that enabled U.S. technologies to migrate overseas. For this reason, we believe that SEMATECH is not the model needed now to achieve U.S. objectives. It is also not clear that an NSF-led program can be structured so it will benefit only U.S. microchip companies. A follow-on study should examine what type of organization should lead this R&D effort.

Exactly what conditions Huawei should be required to meet to regain access to the 5G microchip supply chain might require significant changes to Huawei equipment and could be the subject of future research.

Interagency coordination will be needed to ensure the success of current U.S. 5G activities and the additional activities recommended in Table S.1. The actions of the White House Office of Science and Technology Policy, the FCC, DOC, the Department of Homeland Security, DoD, NSF, NIST, the USPTO, and NTIA will have to be coordinated. Perhaps one organization, such as NTIA or the Office of Science and Technology Policy, should take the lead in the interagency coordination process. A future study could identify options for an effective interagency 5G strategy implementation plan.

Contents

Figures and Tables

Figures

Tables

CHAPTER ONE

Introduction

Once sufficient spectrum is made available to fifth-generation (5G) cellular technologies, 5G networks will provide more capability than fourth-generation (4G) systems. 5G will increase the speed (data rate), efficiency, latency, and size of data transfers (Qualcomm, undated). 5G will facilitate new technologies, services, and systems, such as smart cities, autonomous vehicles, telemedicine, augmented reality, and holographic communications (Medin and Louie, 2019, p. 2; Vodafone, 2018). 5G will provide a network foundation for new services that might provide immense economic benefits to consumers and companies that dominate these new markets (Lewis, 2018, p. 1). However, although 5G promises many new opportunities for U.S. consumers and the U.S. economy, it also presents new risks and vulnerabilities.

This report describes ways in which the United States' adversaries can exploit 5G networks to target the United States and its allies. Adversaries might try to access U.S. 5G networks to compromise U.S. critical infrastructure; spy on the U.S. government; engage in intellectual property (IP) theft against U.S. defense industrial base firms, universities, and high-technology companies; or steal the personal data of U.S. citizens. 5G networks in allied countries could present similar vulnerabilities to allied governments, critical infrastructure, and companies.

Adversary covert access or control of 5G networks, even if only for a short time, could reduce the reliability of U.S. and allied telecommunication networks and the other critical infrastructure that will rely on 5G. These security risks and vulnerabilities pose a growing challenge for the United States, particularly as Chinese companies emerge as major players in 5G markets for 5G infrastructure and user devices.

Objectives

The objective of this report is to identify a comprehensive set of actions the U.S. government can take to secure 5G networks in the United States and allied countries, reduce the risk of compromise of U.S. 5G networks, and ensure that the United States and its allies retain access to trusted sources in the 5G supply chain for 5G infrastructure, systems, mobile phones, and microchips.

The United States should consider the possibility that China and its major domestic 5G suppliers could become dominant players in 5G and sixth-generation (6G) supplier markets in the future.[1] Chinese firms have increased their share of the cellular telecommunication infrastructure market. Huawei and ZTE, both Chinese companies with close relationships with the Chinese government, are leading global suppliers of 4G and 5G wireless cellular infrastructure, a market in which U.S. companies no longer have a presence.

Apple is a leading provider of 4G handsets, and both Apple and Google dominate the market for cellular smartphone operating systems (OSs). However, Chinese mobile phone makers have emerged as major competitors in the smartphone market, and Huawei, even while suffering setbacks because of U.S. sanctions, is

---

[1] 6G refers to the sixth generation of wireless telecommunication networking that will follow 5G. Research and development (R&D) on 6G was formally started by standard-setting organizations (SSOs) in 2021.

challenging Apple and Google in the OS market. U.S. high-tech firms currently dominate the global market for many 4G applications, but Chinese companies, such as Tencent, Baidu, and Alibaba, have also become major players in the mobile-application and related markets for internet messaging, e-commerce, and new types of financial services.

Because Chinese firms could someday obtain dominant market positions in each segment of the 5G market, the United States should develop an understanding of the security implications of China's role in 5G.

The White House's 5G National Security Strategy

National Strategy to Secure 5G of the United States of America (White House, 2020) acknowledges the risks of 5G from malicious actors seeking to exploit the technology for nefarious purposes. The security strategy provides four lines of effort to ensure that the United States deploys and manages a secure 5G infrastructure:

- Facilitate the rollout of 5G in the United States.
- Assess the risks and identifying core security principles for 5G infrastructure.
- Manage the risks to U.S. economic and national security from the use of 5G infrastructure.
- Promote responsible global development and deployment of 5G infrastructure (White House, 2020).

Although the 5G security strategy provides these overarching lines of effort consistent with the threat posed by bad actors and 5G, it does not provide a clear way to implement these goals and secure 5G for the United States. In other words, the security strategy needs to be followed by detailed concrete steps.

A National Strategy to Secure 5G Implementation Plan

Two government agencies, the National Telecommunications and Information Administration (NTIA) and the Cybersecurity and Infrastructure Security Agency (CISA), have developed plans and strategies in support of the White House 5G security strategy. First, NTIA has developed an implementation plan to address the White House security strategy lines of effort (NTIA, 2021). The NTIA implementation plan lists activities the federal government and others plan to carry out to implement the 5G security strategy. Many of these activities are described in high-level terms. In this report, we delve into details in several key areas and suggest specific actions and additional steps U.S. government agencies can take to achieve the goals laid out in the 5G security strategy.

The Cybersecurity and Infrastructure Security Agency 5G Strategy

CISA has developed a 5G strategy focused on security and resilience. CISA's strategic initiatives are as follows:

- Support development of 5G policy and standards by emphasizing security and resilience.
- Expand situational awareness of 5G supply chain risks, and promote security measures.
- Partner with stakeholders to strengthen and secure existing infrastructure to support future 5G deployments.
- Encourage innovation in the 5G marketplace to foster trusted 5G vendors.
- Analyze potential 5G use cases and share information on identified risk management strategies (CISA, 2020).

The CISA strategy builds on the 5G security strategy's lines of effort to propose actions the United States might pursue to secure U.S. 5G infrastructure. In this report, we examine several issues related to many of the initiative areas in the CISA strategy and recommend ways these strategic initiatives can be accomplished.

Methods

We employed a variety of methods in this research. We examined the organizational structure, composition, and work plan for the SSO responsible for the development of 5G standards and architecture. This included an examination of 5G technical standards and standard-essential patents (SEPs) for key parts of the 5G architecture. As part of this analysis, we obtained a list of key technical standards for the 5G radio access network (RAN) and examined patent claims of selected companies associated with 5G technical standards. We next identified IP claims of selected companies for specific 5G technical standards, using the European Technical Standards Institute (ETSI) Intellectual Property Rights (IPR) database (ETSI, undated).

We also conducted a literature survey on cellular technology–related patents and patent infringement cases, patent application filing behavior in China and the United States and alleged IP theft from U.S. patent application filings that are available online. We conducted a literature survey on the alleged espionage, IP theft, and sanction violations by Chinese cellular equipment suppliers, including, in particular, Huawei's behavior in global markets. We examined the reported suspicious behavior of Chinese telecommunication service providers in the United States. Finally, we analyzed current U.S. government 5G policy, strategy, and implementation plans. Finally, we conducted a market analysis of 5G phones, infrastructure, and microchips.

Report Outline

The report is framed in terms of the 5G security competition between the United States and China. It proceeds in five parts. Chapter Two describes China's ascendance in telecommunication markets and the evolution of the telecommunication industry. Chapter Three describes key elements of the competition in 5G security between China and the United States. Chapter Four describes the security competition in the 5G supply chain. Chapter Five outlines potential risks and vulnerabilities of 5G and how China could exploit these to achieve its economic and national security goals. Chapter Six proposes several courses of action the United States can pursue to mitigate the risks posed by the 5G security competition.

CHAPTER TWO

The Rise of Chinese Firms in Global Telecommunication Markets

In this chapter, we briefly review global telecommunication markets and the rise of Chinese firms in these markets.

The Structure of the Global Cellular Infrastructure Market

Cellular communications have advanced rapidly in the past several decades. After competing standards were developed independently in different countries, an international organization called the 3rd Generation Partnership Project (3GPP), was established to set global standards for cellular communications. 3GPP consists of seven regional SSOs that are responsible for setting technical standards for information and communication technology (ICT).[1] A full member of 3GPP must be a member of one of the seven regional SSOs. Members of the regional SSOs are typically drawn from companies developing technologies and systems for cellular networks or from cellular network operators. 3GPP grew out of past conflicts between companies and regional SSOs to set technical standards for second- and third-generation (3G) cellular networks. In these political and technical battles, different technologies by U.S. and European companies were proposed for adoption as global technical standards. 3GPP emerged as the single global SSO able to set such standards (Baron and Gupta, 2018). The advantage of adopting global standards is that it enables mobile devices, such as smartphones, to be interoperable with cellular networks anywhere in the world. 3GPP standards enable users to use or roam on cellular networks anywhere on earth. Another advantage of international standards is that they enable the largest possible economies of scale in the mass production of mobile phones and other components. Today, 5.16 billion mobile phones are operated globally (DataReportal, undated). Many of these phones, given a full-roaming rate plan, can access virtually any cellular network in the world.

The companies leading the development of cellular infrastructure have changed over time. Table 2.1 shows where the leading firms are domiciled and how their market share has changed over time from the deployment of 4G in 2014 through 5G in 2020. In the past five years, Ericsson, Huawei, and Nokia have traded the top market share spot in the cellular infrastructure market.

In the past decade, there has been further consolidation in the cellular infrastructure business. In 2015, Alcatel-Lucent was acquired by Nokia in an effort to better compete with Ericsson and its Chinese rivals (Griffith, 2015). Despite fierce competition, in the past decade, Huawei has maintained a leading position in the market and, in the past two years, has become the market leader.

One of the principal suppliers to the U.S. market, Nokia, based in Finland, has struggled to compete against heavily subsidized rivals, such as Huawei. In 2018 and again in March 2020, it took out substantial loans from the European Investment Bank (EIB) (A. Morris, 2018). To conserve cash, Nokia halted dividend

---

[1] See 3GPP, undated i, for more information.

TABLE 2.1

Major Cellular Infrastructure Vendors and Their Estimated Market Shares

Company	Country	Market Share, as a Percentage			
		2014	2016	2018	2020
Alcatel-Lucent	France–United States	6	0	0	0
Ericsson	Sweden	31	18	27	29
Huawei	China	39	22	31	31
Nokia	Finland	19	32	22	23
Samsung	South Korea	<5	8	11	4
ZTE	China	<5	12	5	9

SOURCES: Indrayan, 2014; Lewis, 2018; Omdia, 2019; TelecomLead.com, 2016.

NOTE: Here we are referring to cellular infrastructure vendors and not vendors for data network components, such as internet routers and switches. However, the 2020 market share figures probably do include vendors that provide 5G edge computing nodes.

payments in late 2019 and, as of the third quarter of 2021, had not reinstated a dividend. The company has said that it needs to invest more than previously anticipated for R&D to develop competitive 5G products, which is likely why it asked the EIB for a $561 million loan in 2020 (Virki, 2020). Nokia reportedly received a takeover offer shortly after the EIB loan was announced (Kapko, 2020). However, it has remained independent, and its finances have recently improved as U.S. sanctions against Huawei and concerns about the cybersecurity of Huawei 5G products have led to market share gains in some countries.

Ericsson, another European supplier to U.S. cellular carriers, is in better financial shape than Nokia, but it has also suffered financial stress. It experienced periodic losses in 2018 and 2019, but, after years of R&D spending on 5G technologies, its fortunes might improve as wireless carriers around the world start upgrading their networks to 5G.

Huawei is a private Chinese company and does not report detailed financial results, although it did report revenue of $122 billion in 2019, an 18-percent increase from 2018 (Strumpf, 2019b). Huawei's ownership structure is murky. Little is available publicly on who its owners are and whether state-owned companies hold a dominant share in the firm. On its website, Huawei claims that it is wholly owned by its shareholders and that no government agency or outside organization holds shares in it (Huawei, undated).

ZTE is a Chinese company that is listed on the Hong Kong and Shenzhen stock exchanges. It also trades on the over-the-counter market as an American depositary receipt in the United States (as ZTCOY). More is known about ZTE's ownership structure than about Huawei's. In early 2019, at least one Chinese state–owned company held a 30-percent ownership stake in ZTE, although it is likely that this stake was reduced to 15 percent later in 2019. ZTE's state-backed parent company reportedly held a much larger stake in the company in 2004, with a 44-percent holding (Strumpf, 2019a). The ZTE ownership history indicates that cellular communication has been a long-term priority for the Chinese government since the company's founding in 1985.

For decades, the U.S. government has let free-market forces operate unencumbered in the cellular telecommunication market. This stance seems to have served the country well in that U.S. companies maintain a significant presence in markets for mobile phones, mobile software and applications, and microchip components. However, as Table 2.1 shows, there is no longer a U.S. "prime contractor" in the mobile infrastructure market. The companies providing 5G mobile network infrastructure in the United States are now primarily European and Chinese companies, which have significant market share globally outside of the United States.

Has U.S. policy emphasis on open markets overlooked other risks that could affect U.S. national security at home and abroad? China's view of the cellular telecommunication industry differs from that of the United States. In this and the next chapter, we argue that China sees all 5G markets, including the 5G RAN, core network, and mobile device markets (which are defined in the next chapter) as key markets in which Chinese companies must compete. China believes that, by having a robust presence in these markets, it can accomplish important national security, foreign policy, and intelligence collection objectives.

The Rise of Huawei and ZTE

China grew its domestic telecommunication industry by subsidizing Chinese companies and stealing IP from other countries. In the early 2000s, Huawei stole technological IP from the United States, and there are credible allegations that IP was also stolen from Canada's leading telecom company, Nortel (Pearson, 2020). In 2003, Huawei was convicted of stealing software code from Cisco Systems, a U.S. company, and, in January 2019, the U.S. Department of Justice indicted Huawei for the theft of trade secrets, wire fraud, and obstruction of justice (Office of Public Affairs, 2019; Thurm, 2003). Internally, Huawei has also implemented financial incentives—providing cash bonuses—to employees who steal IP from competitors (Clancy, 2019).

China has also been accused of using state subsidies and unfair trade practices to grow its domestic telecommunication industry (Clancy, 2019; Clark, 2020). By some accounts, support from China Development Bank allows Huawei to make bids 40 percent lower than competitors for telecom projects (Clancy, 2019). In late 2019, the *Wall Street Journal* reported that Huawei has received more than $75 billion in subsidies of various kinds from the Chinese government since its founding (Yap, 2019). In addition to tax breaks, special deals on real estate, and other subsidies, the *Journal* estimated that, since 1998, Huawei received $16 billion in loans, export credits, and other forms of financing from Chinese banks for itself or its customers, a figure that dwarfs the state support received by its European competitors and that reportedly enabled Huawei to undercut the prices it offered wireless carriers by as much as 30 percent (Yap, 2019).

The combination of stolen IP provided by cyber-hacking campaigns and large financial subsidies from the state and Chinese banks has enabled Chinese cellular infrastructure companies to become major players in 5G markets in more than 170 countries—but not in the United States because of policy restrictions on the use of Huawei and ZTE equipment in U.S. cellular networks.

As 5G networks are being deployed in the United States and other countries, the concern is not that U.S. carriers will turn to Chinese companies to deploy 5G. Rather, it is that this global rollout of the cellular infrastructure providers on which the United States relies could become weakened financially. If they lose market share in other global markets because Huawei and ZTE can offer 5G equipment at lower prices—a probable benefit of the state subsidies in their business models—the financial health of Nokia, Ericsson, and Samsung could decline. If some of these foreign suppliers fail, Huawei and ZTE might be able to obtain dominant global market positions in the cellular infrastructure market over the long term, and U.S. suppliers could be forced to exit the market entirely.

China's Exploitation of 4G Telecommunication Networks

China has a malign history of using telecommunications to create and exploit vulnerabilities in networks for espionage. Huawei poses security risks in the 5G equipment it provides from both "front doors" and "back doors." This section first discusses the backdoor security risks posed by Huawei. *Back door* refers to a vulnerability that allows an actor to bypass security controls and access a computer network or encrypted data (Lepido, 2019). Chinese telecommunication companies, such as Huawei, and hackers alleged to have Chinese-state backing have demonstrated the ability to exploit back doors for espionage and data exfiltration.

Compromise of Law Enforcement Access Points

U.S. law requires that telecommunication equipment manufacturers build into their hardware ways for authorities to access networks for lawful purposes. These access points, although intended for lawful purposes, do provide a means for a malicious actor to covertly access a network. Further, although a company might manufacture the hardware used for a network, that company does not operate the network itself. Therefore, to enhance the security and transparency of the network for the network operator, 5G equipment manufacturers should consider developing additional safety measures. These measures would involve building the hardware so that, before an external actor, or the manufacturer itself, can access the network, it must gain consent from the operator. Also, to date, not all cellular infrastructure providers have complied with our suggested approach.

Backdoor Access

In 2020, the United States announced that Huawei could covertly access mobile-phone networks through back doors in system components designed to aid law enforcement investigations (i.e., for so-called wire taps). Intelligence reporting suggests that Huawei had been exploiting this secret capability for more than a decade (Pancevski, 2020). U.S. officials have recently alleged that Huawei continues to violate this requirement (Pancevski, 2020). Further, hackers believed to be backed by the Chinese Communist Party (CCP) advanced persistent threat (APT) called APT-10 (a cyber-espionage group) have infiltrated the cellular networks of at least ten global carriers since roughly 2012 (Martin and Dou, 2019; Nichols, 2019). APT-10 threat actors stole user geolocation data, text-messaging records, and call logs. These threat actors used virtual private networks to exfiltrate large amounts of data from cellular network servers and set up covert command-and-control networks within the telecommunication networks of undisclosed carriers around the world (Martin and Dou, 2019; Nichols, 2019).

Front-Door Access

China has also demonstrated its intent and ability to deploy malicious code on previously secure equipment to conduct espionage and exfiltrate data. Software update capabilities provide a potential front door through which the security status of telecom equipment can be changed from secure to compromised by an untrustworthy telecommunication infrastructure supplier. The term *front door* refers to the initial delivery of equipment in a secure form to a customer and then, through a software update, the installation of malicious code that makes the equipment insecure, unbeknownst to the customer (Clancy, 2019). Equipment can be sold and installed in a perfectly secure condition without back doors or hidden vulnerabilities. The equipment is secure but can be updated with new software from the supplier.

The software update delivered through a front door, or a disclosed software update portal, can contain hidden vulnerabilities and access points. It is increasingly common for software updates to be delivered automatically without customer or network operator intervention when the capability involves real-time critical applications (e.g., cybersecurity applications or telecom backbone networks).

A possible example of China's use of the front-door approach includes the Chinese cyber-espionage campaign against the African Union from 2012 to 2017, using Huawei equipment (Dahir, 2018). In 2006, China provided $200 billion to build the African Union headquarters, leveraging Huawei for the communication infrastructure. The African Union realized in 2017 that, from 2012 to 2017, Huawei had been conducting espionage activities each night from midnight to 2 a.m., transferring data to servers in Shanghai (Vaswani, 2019).

China's National Intelligence Law

In 2017, the Chinese government publicly shifted toward state-directed cyber espionage when President Xi Jinping restructured Chinese intelligence priorities to focus on cyber spying (Lewis, 2019). Xi stated that these new priorities were part of a larger military modernization effort (Lewis, 2019). In another public declaration, China passed its National Intelligence Law in 2017, which requires all Chinese companies to comply with requests for assistance from the Ministry of State Security without any option for an appeal (Lewis, 2019).

The 2017 National Intelligence Law not only outlines China's more-public efforts to encourage cyber espionage but also illustrates the linkage between the CCP and China's telecommunication industry. Moreover, evidence from the African Union example from 2012 to 2017, as well as the 2017 law, suggests that Huawei acts as a proxy of the CCP. The Huawei–CCP relationship demonstrates why China should be viewed as the United States' primary competitor in the 5G security competition. Additionally, it shows why Huawei and Chinese telecommunication companies in general raise important security concerns for 5G that the United States needs to consider. China and Huawei pose particular security risks because of the potential for hidden back doors in the 5G architecture and the use of legitimate front doors for espionage. Additionally, supply chain risks persist and make up a key element of the security competition, discussed in Chapter Three.

Elements of the 5G Security Competition

In the past, competition in U.S. telecommunication markets has been viewed predominantly as a domestic issue, with a focus primarily on economic and technology aspects. The security of U.S. networks was less of a concern than economics and technology because the major players in the market were U.S. or European companies, which were presumed to be trustworthy. It was also assumed that cellular networks could be made secure by the actions and oversight of the wireless carriers in the United States.

In this chapter, we examine the important ways in which the 5G architecture differs from that of earlier generations, how an adversary could exploit these differences, and how difficult it could be for a government or a wireless carrier to secure 5G networks.

5G Architecture

The 5G architecture is the most advanced network yet developed by the global cellular communication R&D community. It has three major physical segments:

- the core network
- the RAN
- user equipment (UE).

The 5G architecture is illustrated in Figure 3.1. It is the first cellular network that will have fully virtualized servers in its core network that can be hosted by cloud service providers (5G Infrastructure Public Private Partnership Architecture Working Group [WG], 2019).

The advantage of operating on a cloud is that 5G core network servers do not require specially designed computer hardware. They can be hosted on general-purpose computing nodes, enabling 5G wireless carriers to host their servers in their own computing clouds or in those provided by commercial cloud service providers, such as Amazon Web Services. The transition away from hardware-defined networking to software-defined networking in 5G enables more-flexible and -dynamic control of the 5G network, as indicated in Figure 3.1.[1]

Although the transition of the 5G core to the cloud brings cost savings and enormous flexibility, it also introduces potential cybersecurity concerns for national governments, 5G network operators, and 5G users. Some of these concerns are not new, especially as they relate to the role of Huawei and ZTE in building cellular infrastructure and providing mobile devices to users worldwide. However, new technical aspects of the 5G architecture present new cybersecurity challenges, which untrusted actors, such as Huawei and ZTE, can exploit.

---

[1] There are many aspects to SDN, including, in 5G, providing the ability to create virtual networks for different customers at different service levels and to enhance network services and bandwidth dynamically by "spinning up" additional network servers as needed in the core cloud.

FIGURE 3.1

Example of the 5G Architecture for Vehicle Automation

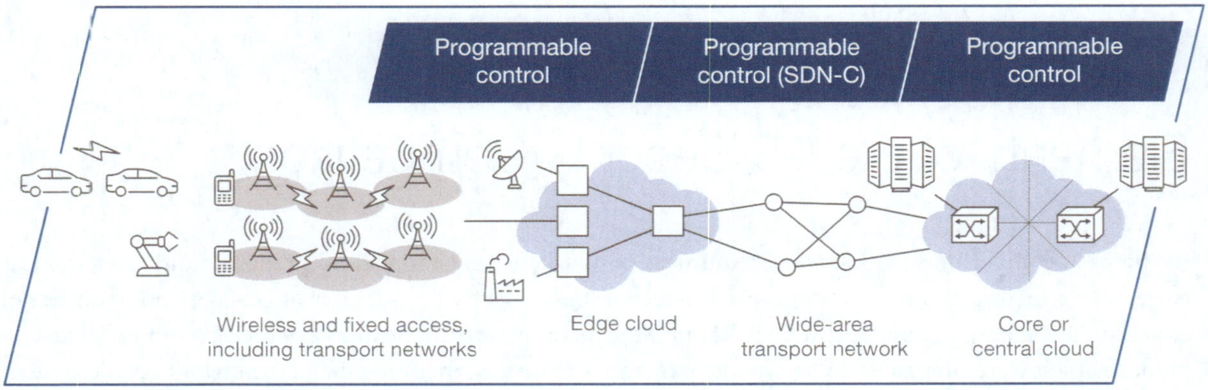

SOURCE: 5G Infrastructure Public Private Partnership Architecture WG, 2019, Figure 2-2. Used with permission.
NOTE: SDN-C = software-defined network controller.

The 5G RAN is composed of much more-flexible base stations, called gNodeBs, that can host edge computing nodes or so-called *edge cloud nodes*, also depicted in Figure 3.1. These will process data that users request much closer to the edge of the network. Edge clouds will support local data storage by content delivery services in order to minimize traffic flow in the wide-area transport network.

Expanded Set of 5G Use Cases

5G will support a much wider range of potential use cases, both for new types of low-latency, high-bandwidth applications for human users and for new types of machine-to-machine communications to support the internet of things (IoT). 5G will support more communication links to people and IoT devices than any previous type of cellular network has (Bonds et al., 2021). 5G is designed to support a very high density of machine-to-machine links and links to UE—up to 10 million devices per square kilometer (National Academies of Sciences, Engineering, and Medicine, 2019). Future 5G UE will include not only current UE, such as mobile phones, tablets, and smart watches, but might also include new UE types, such as eyeglasses, rings, and bracelets, that track a user's vital signs and sleep and that might also provide other capabilities. Future 5G-connected IoT devices will include manufacturing robots, warehouse robots and automated inventory tracking systems, connected vehicles, autonomous vehicles (as shown in Figure 3.1), power plant monitoring and control systems, and other types of building control and industrial control systems, all parts of the expanding IoT ecosystem (Cisco, 2019).

5G Technical Standards

Technical standards for the 5G architecture are developed by the international SSO 3GPP. These standards ensure the interoperability of cellular communication networks and end user devices, such as mobile phones, so a mobile phone sold in the United States can operate, assuming that roaming agreements exist between carriers, on any cellular network in the world. In this section, we describe the 3GPP, its membership, and how it establishes technical standards. We also describe the role of patents and company IP in defining technical standards. Companies compete and cooperate at the same time in 3GPP when setting technical standards. In the past, U.S. policymakers have let private U.S. companies take the lead in setting technical standards in international SSOs, such as 3GPP.

Concerns have grown over the role of Chinese companies and representatives in international SSOs—in particular, in 3GPP. China is putting in place a policy called China Standards 2035, with the objective of increasing China's role and influence in technical standard–setting processes conducted by SSOs (Gargeyas, 2021). This has led the U.S. Department of Defense (DoD) and NTIA to initiate efforts for the United States to retain or, in some cases, regain influence in SSOs, to ensure that technical standards are established using the best technological approaches, rather than to further the economic or intelligence objectives of a particular country.

The 3rd Generation Partnership Project

The 5G architecture is under development by 3GPP and the companies that belong to and contribute technical experts to it.[2] 3GPP consists of representatives from seven technical SSOs having authority in different regions of the world and other selected organizations.[3] Technical standards developed by 3GPP for the 5G architecture become global standards and enable 5G system vendors to develop products that can be sold in any part of the world and will interoperate with each other, regardless of which company makes the mobile phone or base station. At least in principle, 3GPP standards also enable infrastructure equipment built by different vendors to interoperate and be integrated into a single 5G network.[4]

Different 3GPP WGs develop technical standards for parts of the 5G architecture. 3GPP delegates serving on these WGs are technical experts in the technologies relevant to specific parts of the 5G architecture and would likely represent the interests of the companies for which they work. Our review of 3GPP members—discussed in detail next—indicates that most are company employees from 5G infrastructure or component vendors or of cellular carriers, such as AT&T in the United States. There do not appear to be any government officials designated as full 3GPP members. However, government representatives are sometimes invited to present government views on specific technical or network capability issues to 3GPP WGs. So there are mechanisms for governments to influence 3GPP proceedings and technical developments.

U.S. members of 3GPP come from private U.S. companies, such as Qualcomm, Intel, or AT&T. China also has 3GPP members from Chinese companies. However, some Chinese companies that have 3GPP representatives have been assessed to be state-owned enterprises, such as China Telecom. The three major cellular telecommunication carriers in China are all assessed to be state-owned (Permanent Subcommittee on Investigations, 2020). These three companies, as well as Huawei and ZTE, all have company representatives who serve as full members of 3GPP, including in key leadership positions in 3GPP. Chinese government influence might extend to the topics and views these company representatives put forward in 3GPP meetings and technical deliberations. The close alignment of the Chinese telecommunication industry with the Chinese government is in stark contrast to the substantial separation of government and private-sector roles in the United States and many other Western countries. This separation could provide an advantage to the Chinese government and enable it to influence the deliberations and perhaps even the design of the 5G technical architecture. The U.S. government might need to perform independent technical assessments of the 5G network architecture and, in particular, of its security components before 3GPP finalizes them to detect and prevent undue Chinese government influence on the design of the 5G network architecture.

---

[2] For an overview, see 3GPP, undated a.

[3] For example, the regional SSO for North America is the Alliance for Telecommunications Industry Solutions.

[4] However, this type of core network interoperability is more difficult to achieve than base station–user equipment interoperability.

5G technical standards are divided into three areas: RANs, services and system aspects, and core network and terminals. Figure 3.2 provides a snapshot of the organizational structure of these groups and the WGs.

Leadership for each technical specification group (TSG) varies, with each having a chair and several vice chairs. For the TSG RAN, a representative from Nokia (Finland) serves as the chair, with the three vice chairs from China Mobile (China), Ericsson (Sweden), and NTT DOCOMO (Japan) (3GPP, undated h). For the WGs, leadership by company varies significantly. The chairs for the WGs come from the following companies: Qualcomm, MediaTek Beijing, Ericsson, Futurewei, Motorola Mobility España, and Nokia. Thus, for RANs, Chinese companies chair two WGs, U.S. companies chair one, and European companies chair three. Each WG also has one or two vice chairs. For the RAN WG, Chinese companies serve in three vice chair roles, U.S. companies in two, Japanese companies in three, and European companies in two (3GPP, undated b; 3GPP, undated c; 3GPP, undated d; 3GPP, undated e; 3GPP, undated f; 3GPP, undated g). Although the leadership structure and country representation for the RAN WG does not suggest dominance for any particular country, it does show that the United States holds the minority of intellectual leadership positions in the RAN 3GPP WG structure.

3GPP WGs endeavor to develop 5G technical standards using a consensus-building process based on the technical merits of proposals for specific standards. In this process, WG members submit candidate technologies and IP from their own companies. The technical WGs deliberate on proposals brought by competing companies to the WGs for technical standards that will eventually specify technical solutions for algorithms, subsystems, features, or functions of the 5G architecture. If a consensus cannot be reached on a member proposal for a new technical standard, the WG chair must call for a vote to select among competing proposed technical standards. A technical standard must be approved by at least 71 percent of the WG's members to be accepted as an approved technical standard (Baron and Gupta, 2018).

Specific technologies, IP, and patents owned by competing companies often lead to differing agendas and alliances in 3GPP WGs. In the past, the U.S. government generally stayed out of these debates and let private

FIGURE 3.2
3GPP Technical Specification Groups and Working Groups

TSG RAN	TSG services and system aspects	TSG core network and terminals
RAN WG1	SA WG1	CT WG1
RAN WG2	SA WG2	CT WG2
RAN WG3	SA WG3	CT WG3
RAN WG4	SA WG4	CT WG4
RAN WG5	SA WG5	CT WG5
RAN WG6	SA WG6	
RAN Ad Hoc 1		

SOURCE: 3GPP, undated m.

U.S. companies pursue their own agendas in the 3GPP WGs and in the selection of cellular network technical standards in general (Nunno, 2003). However, Chinese firms have increased their presence and activity in 3GPP WGs. In addition, the Chinese government has allegedly pressured Chinese company representatives to vote for technical standards proposed by Chinese companies, regardless of their technical merit, in international standard–setting meetings (WG on Trust and Security in 5G Networks, 2021). As a consequence, the U.S. government has made U.S. leadership in development of 5G standards a priority (NTIA, 2021, Annex D).

Standard-Essential Patents

The IP of 3GPP member companies can become essential ingredients to 3GPP standards for 5G. As mentioned above, member companies compete with each other to have their IP designated as essential for a particular standard. These companies protect, or attempt to protect, their IP by patenting their inventions. These patents are then associated with particular technical standards, as candidate or declared SEPs. A large number of competing patents may be submitted to 3GPP as SEPs, but only a small fraction are typically accepted as SEPs. This has led to some patent-related issues, which we briefly describe in this section.

If specific algorithms or technologies, and the patents associated with them, are deemed to be essential to a 5G technical standard, the contributing company can declare these patents to be SEPs. In some cases, it might not be possible to build a component that complies with a 3GPP technical standard without using a SEP.

As stated in 3GPP working procedures, members must make their declared SEPs available to all companies at fair, reasonable, and nondiscriminatory royalty rates (3GPP, 2021a). However, patent infringement and royalty cases have been found to be rare for approved SEPs, at least in U.S. court cases, and, in fact, most patent infringement cases that have been brought to court in the United States in the mobile phone industry have not involved SEP, but instead have centered on other patents (Taffet, 2016). Richard Taffet, a partner at the law firm Morgan Lewis and an experienced litigator in IP rights, has argued that approved SEPs have not hindered innovation and that most patent infringement cases have involved nonessential patents for cellular phone systems (Taffet, 2016).

Why should anyone be concerned about IP claims and SEPs when considering 5G security issues? One way a SEP holder can offer its IP to a competing company or to a 5G component maker is to sell a component that contains or uses the IP, such as a microchip, rather than divulge IP or sell an IP license to a competing company. Some companies price their IP licenses at high levels but sell their microchips, which contain their IP, at reasonable prices. Microchips are complex and today contain billions of transistors and must be trustworthy. Microchips can be implanted with undocumented circuits and algorithms that might have hidden back doors. In this way, the manipulation of SEPs and microchip pricing and availability can become important levers in the 5G security competition between the United States and China.

5G Radio Access Network Technical Standards and Standard-Essential Patents

SEPs are declared by the 3GPP member companies that have offered their IP to define 5G technical standards. SEPs can be declared before a patent-granting entity agrees to grant a patent for the application and before 3GPP approves the technology in the patent application as essential. There are hundreds of such 5G technical standards and SEPs. We examined patents for the first two layers of RAN technical standards and the SEPs associated with them.

Patents for declared SEPs can be filed in different jurisdictions, including the U.S. Patent and Trademark Office (USPTO), the European Patent Office (EPO), the Japan Patent Office, or the China National Intellectual Property Administration (CNIPA). It is possible for conflicting declared SEPs to be filed in different jurisdictions. When an individual patent jurisdiction's entity grants a patent associated with a declared SEP, the entity does not designate the patent as a 5G SEP—only that the IP claim is valid in that particular jurisdiction. In this analysis, we considered only SEPs, not approved patents.

Radio Access Network Layer 1

Layer 1 establishes technical standards for the RAN physical layer, including standards for channel modulation, coding, and multiplexing schemes. The technical standards for 5G RAN layer 1 are shown in Table 3.1. Also shown in this table are the SEPs that have been established for these technical standards and the organizations that own these declared SEPs.

Surprisingly, as shown in Table 3.1, relatively few SEPs have been granted for layer 1, even though the technologies and technical standards for layer 1 are complex. The only U.S. company granted any SEPs (two)

TABLE 3.1

5G Radio Access Network Layer 1 Standard-Essential Patents Granted by 3GPP

Technical Standard Number	Title	Number of SEPs	SEP Holder			Current Version
			Name	Country	Number of SEPs	
37.213	Physical layer procedures for shared spectrum channel access	4[a]	Samsung	South Korea	3	16
			Ericsson	Sweden	1	
37.885	Study on evaluation methodology of new V2X use cases for LTE and NR	0		—		15.3
37.985	Overall description of RAN aspects for V2X based on LTE and NR	0		—		1
38.201	NR; physical layer; general description	10[a]	Nokia (5)	Finland	5	16
			Intel	United States	2	
			DOCOMO	Japan	1	
			Blackberry	Canada	1	
			Alcatel-Lucent	France–United States	1	
38.202	NR; services provided by the physical layer	1[a]	ASUSTek	Taiwan	1	16
38.211	NR; physical channels and modulation	1[a]	ASUSTek	Taiwan	1	16
38.212	NR; multiplexing and channel coding	1[a]	ASUSTek	Taiwan	1	16
38.213	NR; physical layer procedures for control	1[a]	ASUSTek	Taiwan	1	16
38.214	NR; physical layer procedures for data	2[a]	ASUSTek	Taiwan	1	16
			Polaran	Turkey	1	
38.215	NR; physical layer measurements	1[a]	ASUSTek	Taiwan	1	16
38.202	Study on new radio access technology, physical layer aspects	0		—		14.2

Table 3.1—Continued

Technical Standard Number	Title	Number of SEPs	SEP Holder			
			Name	Country	Number of SEPs	Current Version
38.812	Study on nonorthogonal multiple access for NR	0		—		16
38.824	Study on physical layer enhancements for NR URLLC	0		—		16
38.840	Study on UE power saving in NR	0		—		16
38.855	Study on NR positioning support	0		—		16
38.866	Study on remote interference management for NR	0		—		16
38.885	Study on NR V2X	0		—		16
38.889	Study on NR-based access to the unlicensed spectrum	0		—		16
38.900	Study on a channel model for the frequency spectrum above 6 GHz	0		—		15
38.901	Study on a channel model for frequencies from 0.5 to 100 GHz	0		—		16

SOURCE: 3GPP, undated k; 3GPP, 2021b.

NOTES: V2X = vehicle to everything. LTE = long-term evolution. NR = new radio. URLLC = ultrareliable and low-latency communication. GHz = gigahertz.

[a] Version 15.

in layer 1 is Intel. As of April 2020, no Chinese company had been granted a SEP in layer 1. Also shown in Table 3.1 is the 5G standard release. The standards shown in the table are for release 16 or earlier because release 16 is the latest version of the 5G architecture standard to be frozen (completed) (3GPP, undated l). It is also the first one to provide the full-bandwidth and data-rate URLLC capabilities of 5G.

Release 17 of the 5G standard will have additional capabilities for industrial networks, IoT, RAN and network slicing, cloud computing, satellite network components, unmanned aerial vehicles, and other capabilities. It is now under development but will not be frozen until mid-2022 (3GPP, 2020). Release 17 has been delayed because of the coronavirus disease 2019 (COVID-19) pandemic (3GPP, undated k).

However, many companies have declared layer 1 SEPs in the hopes that 3GPP will approve them. Table 3.2 shows the SEPs declared by four companies—two U.S. companies and two Chinese companies—for RAN layer 1. Following 3GPP regulations, all of these patent applications have been filed with ETSI, where all 5G-related patents should be filed. The table shows that Huawei has filed by far the greatest number of candidate SEPs. For three of the four key technical standards—physical channels and modulation, physical layer procedures for control, and physical layer procedures for data—Huawei has filed almost 2,000 patents, and, for the fourth, multiplexing and channel coding, it has filed more than 2,300.

Of the four companies we considered for layer 1, ZTE, a Chinese company, has declared the second-highest number of patents. Intel has filed the fewest. For some technical standards, Intel has filed 1,000 fewer patents than the Chinese companies have.

Qualcomm, the other U.S. company we considered in this analysis, has filed between 347 and 656 candidate SEPs for each of the four key technical standards of 38.211, 38.212, 38.213, and 38.214 in layer 1. These numbers are, in some cases, thousands lower than the numbers of patents filed by the two Chinese companies, even though Qualcomm is known as a technology leader in this specific area of cellular communications.

TABLE 3.2

5G Radio Access Network Layer 1 Declared Standard-Essential Patents for Huawei, Intel, Qualcomm, and ZTE

Technical Standard Number	Title	Huawei	Intel	Qualcomm	ZTE	Current Version
37.213	Physical layer procedures for shared spectrum channel access	2			239	16
37.885	Study on an evaluation methodology of new V2X use cases for LTE and NR					15.3
37.985	Overall description of RAN aspects for V2X based on LTE and NR					1
38.201	NR; physical layer; general description		3	3		16
38.202	NR; services provided by the physical layer			1		16
38.202	Study on new radio access technology, physical layer aspects					14.2
38.211	NR; physical channels and modulation	1,876	37	347	1,164	16
38.212	NR; multiplexing and channel coding	2,308	86	299	769	16
38.213	NR; physical layer procedures for control	1,912	125	656	1,292	16
38.214	NR; physical layer procedures for data	1,910	75	466	750	16
38.215	NR; physical layer measurements	150	7	9	5	16
38.812	Study on nonorthogonal multiple access for NR					16
38.824	Study on physical layer enhancements for NR URLLC					16
38.840	Study on UE power saving in NR					16
38.855	Study on NR positioning support					16
38.866	Study on remote interference management for NR					16
38.885	Study on NR V2X					16
38.889	Study on NR-based access to the unlicensed spectrum					16
38.900	Study on a channel model for the frequency spectrum above 6 GHz					15
38.901	Study on a channel model for frequencies from 0.5 to 100 GHz					16

SOURCES: 3GPP, undated b; ETSI, undated. The data shown in this table are the results of extensive analysis, as explained in the "Method" section of Chapter One.

Some other important aspects of these patent declarations should be mentioned here. The patents filed by Intel and Qualcomm are all for specific 5G releases of these technical standards (most are for release 16). In addition, the majority of the patents filed by these U.S. companies are filed in the United States or are filed in multiple countries in accordance with Patent Cooperation Treaty, 2001. In contrast, the majority of patents filed by the two Chinese companies are filed in China alone, although a relatively small number are filed in accordance with the Patent Cooperation Treaty. This patenting strategy by Huawei and ZTE might have little value in the United States or European countries, but it could enable Huawei and ZTE to erect barriers to competition from Western firms in China. For example, it could prove difficult for Qualcomm to offer its chips to mobile phone makers for phones designed for the Chinese market.

The patents shown in Table 3.2 demonstrate the competition to define IP rights for the 5G architecture but are not necessarily indicative of the influence or contribution of individual companies to the standards in question. It is not clear whether 3GPP will approve any of the declared SEPs to these companies in RAN layer 1. It is also not clear, without delving into the technical details of the standard and patents, how many of these patents are overlapping or conflicting or, even more importantly, of low quality. However, given the large number of patents filed, we believe that it is probable that a significant number of patents filed by these Chinese and U.S. companies overlap and present conflicting IP claims.

The way in which these overlapping claims are sorted out via 3GPP's approval processes could have significant implications for the United States' intellectual and technological leadership in 5G, the economic health of the U.S. telecommunication sector, and U.S. national security. One possibility, as mentioned above, is that Chinese companies enforce the claims only in China, forcing Western 5G providers to use Huawei or ZTE algorithms or chipsets within their system offerings in China. Or it is possible that Huawei will succeed in having 3GPP approve a large number of its declared SEPs as 5G SEPs. In this case, non-Chinese companies could be forced to use Huawei algorithms or chipsets in their 5G base stations or mobile phones. This could change the competitive landscape in 5G infrastructure and mobile phone markets and disadvantage U.S. mobile phone manufacturers, such as Apple, and European 5G infrastructure providers, such as Ericsson and Nokia. Nokia told its investors in 2020 that it had to spend much more on R&D to develop competitive 5G infrastructure products, which might indicate that it considers its IP posture deficient relative to Huawei or ZTE. If Huawei or ZTE were to someday hold important SEPs and they decided to make the functionality embodied in these SEPs available to Ericsson and Nokia only as microchips that contained malicious circuitry or back doors implemented in firmware, this could pose a national security threat to the United States and U.S. allies.

The USPTO should monitor the growing competition between U.S. and Chinese firms in the 5G architecture, in SEP submissions, and in 3GPP adjudications of submitted SEP declarations. The USPTO should monitor the unique and overlapping 5G SEPs declared by Chinese and U.S. firms, the dates such patents are filed, where they are filed, SEP quality, and SEP approval, to provide U.S. policymakers indications of whether there is a shift in intellectual leadership in communication technologies from Western to Chinese companies. This information could guide the development of new U.S. government–sponsored research into such technologies, to ensure that U.S. companies retain technical leadership in critical areas of wireless communications.

Radio Access Network Layer 2

RAN layer 2 technical standards define higher-level functions in the 5G RAN—in particular, radio resource control (RRC) logic and algorithms, capabilities that can distinguish efficient and highly capable radio base stations from less capable systems. RAN layer 2 standards define radio interface architecture and protocols, the specification of the RRC protocol, the strategies of radio resource management, and the services provided by the physical layer to the upper layers. The SEPs granted for layer 2 and the companies to which they have been granted are shown in Table 3.3.

Table 3.3 shows that, as of April 2020, relatively few companies had been granted SEPs for RAN layer 2. Major 5G system providers, including Nokia, Ericsson, and Samsung, have SEPs in layer 2. No Chinese company has yet been granted an SEP for RAN layer 2. Intel has been granted one SEP in layer 2 for a patent related to the 37.324 technical standard.

Table 3.4 shows the SEPs declared by the same four companies considered above—Intel, Qualcomm, Huawei, and ZTE—for RAN layer 2. Following 3GPP regulations, all these patent applications for RAN layer 2 have been filed with ETSI. The table shows that Huawei has filed by far the greatest number of declared SEPs. For the technical standard describing the overall structure of layer 2, Huawei has declared 2,560 pat-

TABLE 3.3

5G Radio Access Network Layer 2 Standard-Essential Patents Granted by 3GPP

Technical Standard Number	Title	Number of SEPs	SEP Patent Holder			Current Version
			Name	Country	Number of SEPs	
38.300	NR; overall description; stage 2	0		—		16.0
38.304	NR; UE procedures in idle mode and in an RRC inactive state	1	ASUSTek	Taiwan	1	15.6
38.305	NG-RAN; stage 2 functional specification of UE positioning in NG-RAN	0		—		15.5
38.306	NR; UE radio access capabilities	1[a]	ASUSTek	Taiwan	1	15.8
38.314	NR; layer 2 measurements	1[a]	Ericsson	Sweden	1	Unknown
38.321	NR; MAC protocol specification	3[a]	ASUSTek	Taiwan	3	15.8
38.322	NR; RLC protocol specification	9[a]	Samsung	South Korea	5	15.5
			ASUSTek	Taiwan	1	
			DOCOMO	Japan	1	
			Ericsson	Sweden	1	
			Nokia	Finland	1	
38.323	NR; PDCP specification	14[a]	Samsung	South Korea	10	15.6
			ASUSTek	Taiwan	1	
			DOCOMO	Japan	1	
			HTC	Taiwan	1	
			Nokia	Finland	1	
38.331	NR; RRC protocol specification	3[a]	ASUSTek	Taiwan	3	15.8
38.340	NR; backhaul adaptation protocol	0		—		0.2.0
38.804	Study on NR access technology radio interface protocol aspects	0		—		14.0
38.822	NR; UE feature list	0		—		15.0.1
38.825	Study on NR industrial IoT	0		—		16
38.832	Study on enhancement of RAN slicing for NR	0		—		—
38.836	Study on NR sidelink relay	0		—		—
38.874	NR; study on integrated access and backhaul	0		—		16
37.340	NR; multiconnectivity; overall description; stage 2	0		—		16

Table 3.3—Continued

Technical Standard Number	Title	Number of SEPs	SEP Patent Holder		Number of SEPs	Current Version
			Name	Country		
37.324	E-UTRA and NR; service data adaptation protocol specification	10[a]	Samsung	South Korea	4	15.1
			ASUSTek	Taiwan	1	
			DOCOMO	Japan	1	
			HTC	Taiwan	1	
			Intel	United States	1	
			Nokia	Finland	1	
			Sharp	Japan	1	
37.355	LTE positioning protocol	0	—			15
37.873	Study on optimizations of UE radio capability signaling; NR and E-UTRA network aspects	0	—			16

SOURCE: 3GPP, undated c.

NOTES: NG-RAN = next-generation RAN. MAC = medium access control. RLC = radio link control. PDCP = packet data convergence protocol. E-UTRA = evolved universal terrestrial radio access.

[a] Version 15.

ents to be SEPs, while Intel has filed for just seven patents for this technical standard (38.300) to be SEPs. Qualcomm has declared 124 patents to be SEPs for this same technical standard. Table 3.4 shows a similar pattern of declared SEP filings for each of the four companies. Huawei consistently has filed far more patents and SEPs than any of other companies. ZTE has filed the next-highest number of patents, while the two U.S. companies consistently file fewer patents in each of the technical standards defining layer 2.

Almost all SEP declarations of the four companies are for the same layer 2 technical standards—for example, 38.300 (overall description), 38.321 (MAC protocol), 38.322 (RLC protocol), 38.223 (PDCP), and 38.331 (RRC). The four companies appear to be competing to define the key technologies in these key technical standard areas that form the heart of layer 2.

Competition in Developing Standards

The analysis of declared SEPs for RAN layers 1 and 2 indicates that, even before 5G systems are sold and deployed, and before vendors submit bids for contracts with wireless carriers, competition between 5G system developers was already underway in the 3GPP standard-setting process. The competition for SEPs has not ended but rather is continuing and was not resolved in release 16 for RAN layer 1 and 2 standards. It is possible that 3GPP will not decide among the many competing SEPs until later releases of the 5G standard.

Table 3.5 summarizes the 5G SEP competition for RAN layers 1 and 2. The table shows that Huawei has declared thousands more SEPs than two of the most-important U.S. companies in this part of the 5G architecture, Qualcomm and Intel (in 2019, both were positioning themselves to be deliver components that would implement RAN layer 1 and 2 standards). ZTE has adopted the same strategy as Huawei and has also declared more SEPs in RAN layers 1 and 2 than either Intel or Qualcomm.

We also found that, as of 2020, 3GPP had approved only a few of these declared SEPs. So the vast majority of patents referred to in Table 3.5 are claimed as SEPs only by the IP holders. Different SSOs use different

TABLE 3.4

5G Radio Access Network Layer 2 Declared Standard-Essential Patents for Huawei, Intel, Qualcomm, and ZTE

Technical Standard Number	Title	Huawei	Intel	Qualcomm	ZTE
37.324	E-UTRA and NR; service data adaptation protocol specification	19		1	2
37.340	NR; multiconnectivity; overall description; stage 2	542	2	26	12
37.355	LTE positioning protocol				
37.873	Study on optimizations of UE radio capability signaling; NR and E-UTRA network aspects				
38.300	NR; overall description; stage 2	2,560	7	124	210
38.304	NR; UE procedures in idle mode and in the RRC inactive state	91		14	24
38.305	NG-RAN; stage 2 functional specification of UE positioning in NG-RAN	1			13
38.306	NR; UE radio access capabilities	1		21	8
38.314	NR; layer 2 measurements			1	
38.321	NR; MAC protocol specification	2,014	32	142	635
38.322	NR; RLC protocol specification	1,247	3	14	89
38.323	NR; PDCP specification	511	3	23	45
38.331	NR; RRC; protocol specification	3,030	129	251	1,499
38.340	NR; backhaul adaptation protocol				
38.804	Study on new radio access technology radio interface protocol aspects				
38.822	NR; UE feature list				
38.825	Study on the NR industrial IoT				
38.832	Study on the enhancement of RAN slicing for NR				
38.836	Study on the NR sidelink relay				
38.874	NR; study on integrated access and backhaul				

SOURCES: 3GPP, undated c; ETSI, undated.

TABLE 3.5

Standard-Essential Patents Declared by Huawei, ZTE, Intel, and Qualcomm for 5G Radio Access Network Layers 1 and 2

Company	RAN Layer 1	RAN Layer 2	Total
Huawei	8,058	10,016	18,064
Intel	333	176	509
Qualcomm	1,781	617	2,398
ZTE	4,219	2,537	5,667

SOURCES: 3GPP, undated b; 3GPP, undated c.

definitions of *essentiality*, a legal topic that is beyond the scope of this particular study (Contreras, 2017). In addition, the timing required for SEP disclosure to the SSO and for the SSO to approve or disapprove each SEP declared by each claimant differs a great deal. In other words, a claimant may declare a patent application a SEP without necessarily disclosing the contents of the declared SEP to all parties. In addition, in some cases, the SSO may decide to wait for the patent applications to be validated by the national or regional patent office that decides the validity of patent applications (Contreras, 2017). Given the complicated processes that can be required for 3GPP approval of declared SEPs, it is not surprising that so few 5G SEPs have so far been approved.

Nevertheless, 3GPP guidance is clear:

> Individual Members should declare to their Organizational Partners any IPRs which they believe to be essential, or potentially essential, to any work being conducted within 3GPP. During each 3GPP meeting (TSGs and WGs) a call for IPRs must be made by the Chair. (3GPP, undated j)

When a U.S. company declares SEP for a 3GPP standard, the SEP is likely submitted to the U.S. or European SSOs belonging to 3GPP: the Alliance for Telecommunications Industry Solutions or ETSI, respectively. In contrast, Chinese companies, such as Huawei or ZTE, may declare their SEPs to the China Communications Standards Association. For a regional SSO to accept a declared SEP, a patent application should be filed with the associated patent office. There might be differences in declared SEP filing requirements with these different regional SSOs, but, typically, the declared SEP is filed simultaneously with associated regional or national government patent review offices, such as the USPTO, EPO, or CNIPA.

U.S. Patent Application Delays and Chinese Patent Infringements

The number of patents filed for a particular technical standard is not necessarily an indication of a better, unique, or new technical capability, especially for patents filed in China by Chinese companies. Chinese companies have been accused of abusing the international patent system and using it to acquire IP from Japanese and U.S. companies.

As early as 2005, observers noted that U.S. patent applications were being reviewed by parties in China to quickly develop competing products and to file competing patent claims in China. The Japan Patent Office found that Japanese patent applications were being examined 17,000 times daily by users in China. Critics have stated that foreign pirates and counterfeiters, particularly in China but also elsewhere in Asia, were costing U.S. IP owners more than $50 billion a year because of IP theft from patent applications that had been submitted but not yet approved. Analysts also found that U.S. companies and inventors were not filing court injunctions or requesting other legal remedies from Chinese companies that infringed on their patents in China (Choate, 2005). A variety of reasons have been hypothesized for this behavior by U.S. companies (which have not divulged their reasoning in order to avoid offending China), but a detailed discussion of these issues was beyond the scope of the current study.

Chinese companies have defended themselves against patent infringement claims using a technique some have called the Great Wall of Patents (Choate, 2005). This protection scheme can be described as follows:

1. A Chinese company reviews a U.S. patent application online and determines that the Chinese company can develop a product for Chinese or foreign markets based on the approach outlined in the U.S. patent application.
2. The Chinese company quickly files for a patent in China. According to some analysts, some of these Chinese patent applications include exact copies of figures and similar descriptions of subsystems or components contained in the U.S. patent application.

3. With the Chinese patent in hand, the counterfeiter or pirate is then prepared to contest foreign claims of infringement in Chinese courts. The Chinese court system can be so slow that the Chinese company can conduct business and sell pirated products for years before an injunction or other legal remedy is imposed.

One reason this patent infringement approach works is that it takes so long for the USPTO to approve a U.S. patent application. By international agreement and according to U.S. law, the USPTO must publish a patent application on the internet within 18 months after its filing even when no patent has been issued (Choate, 2005). The 18-month rule is even more onerous for U.S. companies and inventors because of the large backlog of patent applications at the USPTO. In 2005, the backlog was nearly 28 months;[5] in 2021, the average time required for a patent to be issued is 23 months, and the current number of backlogged patent applications is more than 600,000 (USPTO, 2021b). At that time, Choate noted that technical patents required even more time to be decided, with some estimates of more than 40 months for a technical patent to be validated or invalidated.

Today, Chinese companies working on 5G technologies might continue to use the Great Wall of Patents approach to obtain technology know-how and patent claims to prepare their own declared SEPs. At a minimum, Chinese companies can exploit these SEPS in their own domestic markets, but they might also be able to expand to markets in Africa or Asia, where IP protection schemes are weak.

Patent Quality and 5G Standard-Essential Patent Quality

According to the World Intellectual Property Organization (WIPO), more than 3 million patent applications were filed worldwide in 2016, up 8.3 percent from 2015. In 2016, the patent offices in the United States, China, Japan, South Korea, and Europe received 84 percent of the world total. CNIPA received 1.3 million patent applications in 2016—more than the combined total of the other four top patent offices worldwide (the USPTO, the Japan Patent Office, the Korean Intellectual Property Office in South Korea, and the European Patent Office) (Pentheroudakis, undated). Without naming China explicitly, WIPO experts have expressed concerns about the quality of patents applied for and perhaps even validated in different jurisdictions. Patent quality is a major concern because it can call into question the reliability, sustainability, and integrity of the global patent system (Pentheroudakis, undated).

Concerns about patent quality are especially acute for technical standards, such as 3GPP 5G technical standards. According to Chryssoula Pentheroudakis, patent quality represents an essential input factor into the technical standardization process:

> Granting patents of poor quality exacerbates the already complex interaction between the standardization system and the patent system. Too many and/or weak patents, and the complex task of determining their validity in the context of litigation have the potential to tilt the negotiation balance, significantly impact transaction costs, and interrupt rapid implementation and innovation via the standardization process. (Pentheroudakis, undated, pp. 3–4)

ETSI, the SSO responsible for 5G standard-setting activities in the European Union, has established links between the databases of 3GPP and the EPO. These linked databases can be used to help assess patent quality or SEP quality. The EPO has also established broad access to drafts of standards from the Institute of Electrical and Electronics Engineers and International Telecommunication Union. However, to date, other patent

---

[5] There is an exception to the 18-month rule, but the inventor must submit a special application for it to the USPTO. By some estimates, only 10 percent of patent applications submitted to the USPTO are submitted with this secrecy exception.

offices have yet not linked their patent databases with SSOs' databases of standards. The majority of U.S. 5G SEPs, at least those we reviewed for this study, are Patent Cooperation Treaty patent applications and so are simultaneously filed with the EPO and transparently linked to 3GPP standards. However, our patent analysis revealed that many—perhaps the majority—of Chinese SEPs are not filed with a link to 3GPP standards.

According to some patent experts, SEP declaration practices of some SSOs do not include enough information about the *essentiality* of declared SEPs. In other words, the SSO technical standard databases contain an increasing number of patent declarations that are "deemed essential to technical standards by the patent holders without sufficient scrutiny regarding that essentiality" (Pentheroudakis, undated, p. 6).

In 2020, the USPTO became concerned about the growing number of suspect trademark and patent applications that Chinese entities filed in the United States. In response, the USPTO conducted a study to determine the reasons for this development (USPTO, 2021a). The growth in Chinese patent filings in the United States is part of a larger worldwide trend of increased Chinese patent filings. In 2019, CNIPA received 1.5 million patent applications from Chinese entities, accounting for nearly half the patent filings in all jurisdictions (USPTO, 2021a). The growth in Chinese patent activity observed by WIPO in 2016 has been maintained in 2019.

The USPTO found evidence that nonmarket factors encouraged Chinese entities to file an increasing number of patents. The primary factor is the subsidies that the Chinese government and local and provincial governments provided to patent applicants. In some cases, these subsidies are greater than the cost to file a patent application. A company can make money filing a patent, even if it never uses the patent in a commercial activity (USPTO, 2021a). Recently, some provincial and local governments in China have increased subsidies for foreign patent filings, which has led to an increase in filings in the United States, as observed by the USPTO. Huawei and ZTE have likely benefited from these subsidies, as shown in Table 3.5. In addition, with the many 5G declared SEPs filed by these Chinese companies in China, they can employ the Great Wall of Patents technology acquisition strategy described earlier in this section.

Chinese national, provincial, and local government subsidies that encourage patent filings calls into question their commercial value. The USPTO found that the commercial value of patents issued in China is lower than that of those issued in the United States. The USPTO compared the following two measures to reach this conclusion:

- the rate at which domestic inventors file for patent protection overseas
- patent licensing fees.

In 2018, for every 100 domestic applications, Chinese applicants filed only five foreign patent applications. In contrast, U.S. entities filed for 80 foreign patents for every 100 domestic applications. Licensing data provide an even stronger indication that Chinese patents are of low quality (USPTO, 2021a). WIPO's 2020 Global Innovation Index shows that the United States ranks first in IP receipts, while China ranks 44th (Cornell University, Institut Européen d'Administration des Affaires, and WIPO, 2020, Appendix II, p. 239). Both measures indicate that Chinese-issued patents, on average, have much lower commercial value than U.S.-issued patents, and therefore Chinese-issued patents are of much lower quality than U.S.-issued patents.

Research, Development, and Patents

The data presented in the previous section show that Chinese entities might be filing many low-quality patent applications and that some Chinese companies could be using IP gleaned from U.S. patents to file competing claims in their domestic market to erect barriers to foreign competition in China and to further their own technology acquisition agendas. However, this does not mean that Chinese companies are not developing their own technologies and commercializing their own R&D IP.

Figure 3.3 shows R&D expenditures in 2020 for companies that spend the most on R&D globally and for some smaller companies in the 5G ecosystem. Amazon spent the most on R&D—almost $43 billion in 2020. Alphabet was the second-largest R&D spender. Huawei, shown in red, was the third company globally in R&D spending in 2020, spending more than $22 billion on R&D, more than either Apple, Intel, or Qualcomm. R&D spending by ZTE is also shown in the red in the figure. It is a smaller company by revenue and spent $2.15 billion on R&D in 2020. Even so, ZTE is looking to develop and use more of its own IP for its 5G products to reduce the licensing fees it pays to other companies (ZTE, 2021).

The R&D expenditures of Huawei and ZTE suggest that some—and perhaps many—of their declared 5G SEPs might be based on technologies they have developed internally. If Huawei's or ZTE's 5G SEPs are eventually approved by 3GPP, they could have more than just financial implications for 5G vendors, mobile phone makers, and wireless carrier ecosystems. That situation could force U.S. and European infrastructure and mobile phone suppliers, as well as wireless carriers, to adopt Huawei or ZTE systems, algorithms, or chipsets into their products and networks. This could have negative financial implications for U.S. and European 5G companies. It could also have detrimental cybersecurity implications for the United States and its allied partners. Huawei 5G chipsets could be designed to include circuitry that could siphon off message traffic in the network under certain conditions or when certain nonpublic commands are sent to 5G base stations that use compromised microchips. It is not hard to embed and hide malicious circuitry in microchips that contains billions of transistors and thousands of circuits. Microchip security will be a primary concern as 5G technical standards are converted to working microchips.

A U.S. strategy for securing U.S. and allied-country 5G infrastructure should include steps to ensure that the IP of Western 5G technology developers is incorporated into 3GPP SEPs and that Western companies retain technical leadership in critical areas of wireless communications. The United States will need to rebuild a robust R&D base in for some parts of the 5G architecture.

FIGURE 3.3

Research and Development Spending of Key Companies

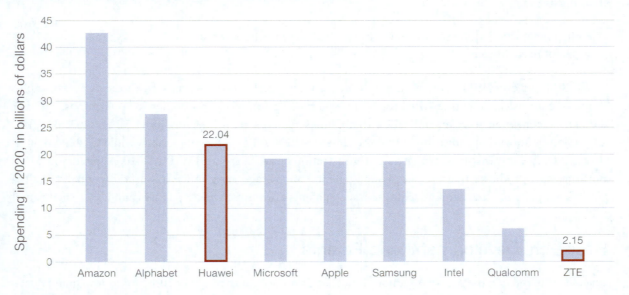

SOURCES: Alsop, 2021; Bajpai, 2021; Intel, 2021; ZTE, 2021.
NOTE: Red outline indicates a company based in China.

The Health of U.S. Telecommunication Research and Development

Before the breakup of AT&T in 1984, the majority of U.S. R&D on telecommunications in the United States was undertaken by Bell Laboratories, which played a key role in the initial development of cellular communications. Another important early mover was Motorola. Motorola was an early player in the cellular infrastructure market, as well as a maker of push-button mobile phones. However, a series of events and competitive market pressures led to spin-offs and the fragmentation of R&D divisions at both companies and a reduction in U.S. corporate R&D spending on wireless telecommunications (Rubin et al., 2021).

The federal government has, at times, funded research in civilian wireless communications but at modest levels (National Research Council, 2006). The U.S. Department of Commerce (DOC) supports a small amount of research in telecommunications. It maintains two laboratories that have, as part of their mission, such research: the Communications Technology Laboratory and the Institute for Telecommunication Sciences (ITS). CTL is operated by the National Institute of Standards and Technology (NIST). ITS is administered by NTIA. In 2015, a National Academy of Sciences study found that both organizations were underfunded and understaffed (National Academies of Sciences, Engineering, and Medicine, 2015). Their funding challenges reflect the low priority that U.S. civilian government agencies place on R&D for civilian wireless communications. The National Science Foundation (NSF) also funds research in telecommunication and wireless communication technologies, but this research portfolio has also been relatively small (National Research Council, 2006). To regain leadership in 5G and in the next generation of civilian wireless communications, including 6G technologies, the United States will need to increase R&D expenditures in this area, as DoD has recently begun to do. Exactly which part of the U.S. government should take the lead in this research—NSF, NTIA, DoD, or some other government agency—and how should it engage and collaborate with U.S. private industry players are important questions but are beyond the scope of this study.

CHAPTER FOUR

Competition in the 5G Supply Chain

The United States–China security competition also occurs within the 5G supply chain, with different actors competing over its various elements. In this chapter, we first describe the major elements of 5G supply chains before discussing the leading companies (in terms of market share) in each sector of the 5G supply chain. We identify where U.S. companies might have a weak position and where they might have a stronger position in the supply chain (i.e., 5G infrastructure and microchip foundries). We also identify where Chinese companies have strong and weak positions in the supply chain. Lastly, we describe the interdependencies between Chinese and U.S. companies in the 5G supply chain and U.S. efforts to hold China accountable for its malign behavior by withholding U.S. technologies from Chinese 5G companies.

5G Supply Chain Elements

Supply chains of the largest 5G companies span the globe. These supply chains can be divided into five major areas:

- 5G infrastructure providers
- microchip designers
- microchip foundries
- mobile phone device providers
- OS providers.

Important dimensions of the competition in the 5G supply chain between China and the United States are economic competition, technology preeminence, and security. We address these dimensions in each of these five areas of the supply chain.

5G Infrastructure Providers

One can think of a 5G infrastructure provider as a prime contractor that builds and sets up 5G networks operated by wireless carriers. As shown in Table 2.1 in Chapter Two, all major 5G infrastructure providers today are foreign companies: Ericsson, Nokia, Samsung, Huawei, and ZTE.

The majority of components that infrastructure providers use to build U.S. 5G networks, base stations, routers, switches, and servers are manufactured or assembled overseas. One of the significant changes in the transition from 4G to 5G is that many of the hardware components that go into the core network will no longer have to be special-purpose equipment and instead can be provided by cloud computing service providers. The unique and new part of 5G core networks will be the software provided by the 5G infrastructure providers. The 5G supply chain for servers, routers, and switches will not be unique to 5G, making it possible for the hardware to be provided by a wide variety of firms. Therefore, the key factor for ensuring the security of 5G core networks is the software provided by the infrastructure providers, not the hardware.

The situation is a bit different at the edge of the 5G network, or the 5G RAN. 5G base stations could have edge computing or edge cloud capabilities, so they will have software-defined components that run on commodity hardware. In addition, 5G base stations will contain unique 5G hardware for the transmission and reception of wireless signals. Each 5G infrastructure provider develops and sells its own 5G base stations. U.S. wireless carriers will be dependent on European infrastructure providers or Samsung for base stations. As noted already, the United States no longer has any companies that make or integrate complete 5G networks, so, given the U.S. government's procurement bans on Huawei and ZTE, U.S. wireless carriers are reliant on Ericsson, Nokia, or Samsung for 5G network infrastructure. These companies have their own supply chains for critical components that go into their base stations.

Today, the United States does not have a direct stake in the economic competition between 5G infrastructure providers, so one view is that the United States should not be concerned about technology preeminence of any provider. However, the United States is dependent on non-Chinese infrastructure providers, and it would be in the U.S. interest if these providers were to supply the United States with the most advanced, secure, and reliable 5G network possible, because having such 5G networks provides a virtual environment that U.S. app developers can use to develop innovative new products.

To ensure the security of the U.S. network, the United States could continue its policy of permitting U.S. wireless carriers to use only non-Chinese infrastructure providers. Alternatively, if it were decided that Chinese 5G equipment was superior technically to that provided by other non-Chinese vendors and this benefit outweighed the risks of a security compromise of U.S. telecommunication networks, the United States could decide to permit Chinese 5G infrastructure in U.S. networks. However, to date, there is no evidence that Chinese 5G infrastructure equipment is superior technically, so the trade-off does not have to be made.

Foreign microchips and software can also be a sources of accidental or intentionally implanted cyber vulnerabilities that can compromise the integrity of ICT. About 80 percent of microchips are now manufactured in Asia, and U.S. microchip supply chains extend into China and other Asian countries (Adee, 2008; Shaffer, 2021). Software is another potential source of vulnerabilities that can alter the functionality, reliability, and integrity of ICT systems. 5G systems are only one example of ICT that can be compromised in this way, but it is nevertheless an important example because 5G networks will become a part of critical infrastructure in many countries, including the United States. Ensuring that compromised microchips and software are not implanted in U.S. 5G networks is a multifaceted challenge.

5G Microchip Designers

5G base stations, servers, and mobile phones contain microchips designed by specialized companies in the supply chains of major 5G system integrators and mobile phone makers. 5G microchips are designed in the United States, Europe, China, or other parts of Asia. Table 4.1 lists the companies that design and, in some cases, manufacture high-value microchips for 5G phones. Most of the microchips designed by U.S. companies are manufactured in Asia. Most U.S. chip designers are *fabless*—that is, they do not own their own foundries (no fabrication). Competition is fierce in the market for high-value chips used in smartphones, including 5G phones. These high-value components include application processors, modems, and radio-frequency front-end (RFFE) microchips. Increasingly leading phone makers, such as Apple, Huawei, and Samsung, have brought design, but not necessarily the manufacturing of the application processors and modems used in their phones, in house, as shown in Table 4.1.

Qualcomm holds a dominant market position for many types of mobile phone components, including application processors and modems. However, for a time, Qualcomm lost market share as phone manufacturers, such as Huawei and Samsung, brought chip development in house (Chin, 2020). The list of chip manufacturers in Table 4.1 does not include Intel, which has invested substantial R&D in SEPs for 5G RAN

TABLE 4.1

Leading Designers of 5G Microchips

Company	Country	Foundry	Chip Type		
			Processor	Modem	RFFE
Apple	United States	No	x		
Qualcomm	United States	No	x	x	x
HiSilicon[a]	China	No	x	x	
MediaTek	Taiwan	No	x	x	
Samsung	South Korea	IDM	x	x	
Murata	Japan	Yes			x
Skyworks	United States	No			x
Qorvo	United States	Yes			x
Broadcom	United States	No			x

NOTE: IDM = integrated device manufacturer (a manufacturer that designs and manufactures its own chips).

[a] HiSilicon is a subsidiary of Huawei.

layers 1 and 2. In the early stages of 5G architecture development, Intel was competing with Huawei, ZTE, and Qualcomm in the race to develop 5G modem chips.

Intel agreed to supply 4G modem chips to Apple when a fairness, reasonableness, and nondiscrimination (FRAND) dispute erupted between Apple and Qualcomm. The two companies went to court, which led to a settlement in April 2019 (Gartenberg, 2019b). As part of the settlement, Apple agreed to pay Qualcomm $4.5 billion in royalty fees. Hours after the settlement was announced, Intel decided to exit the 5G modem business, saying that it would probably not make a profit after having lost Apple as its primary customer (Gartenberg, 2019a). Industry observers speculated at the time that Apple was forced to settle with Qualcomm because Intel was unable to produce 5G modem chips needed for the first Apple 5G-capable iPhone. In hindsight, it appears that this speculation was accurate.

Why was Intel unable to produce 5G modem chips in 2020, when Qualcomm, Samsung, and Huawei all had started production? There are two possible explanations:

- The first is that Intel lacked key IP or declared SEPs in RAN layers 1 and 2. It might have needed to obtain licenses from Huawei, ZTE, Samsung, or Qualcomm for the technologies it lacked to produce 5G modems. The financial terms for these licenses might have been a deterrent to competing in the market. And the political issues associated with licensing technologies from Huawei or ZTE would likely have caused difficulties with U.S. authorities. We view this explanation as unlikely, as noted earlier: Intel does possess significant IP and declared SEPs in RAN layers 1 and 2.
- Another, more plausible explanation is that new 5G modems require state-of-the-art chip speeds, on-chip memory, low power consumption, and the ability to fit in the small form factor of mobile phones. The reason Intel dropped out of the 5G modem competition might have been due to its inability to manufacture 5G microchips. It appears that Intel could not produce the 5G modem chips it designed in its own foundries in 2020 because of delays in perfecting its 10-nanometer (nm) process technology. Intel's most-advanced foundries use a 10-nm process, but the company has struggled to perfect this process in the past decade (Williams, 2018). This led to delays in the introduction of new products—in particular, its main product, computer microprocessors. The delay, in turn, made it possible for Advanced Micro Devices (AMD) to introduce microprocessors with superior performance that are fabricated by Taiwan

Semiconductor Manufacturing Company (TSMC) using its new 7-nm process. AMD has pressed its advantage over Intel in the microprocessor market and was expected to become TSMC's biggest customer in 2020 utilizing its 7-nm foundry (Mayersen, 2020). Intel's difficulty in producing 10-nm chips might have prevented Intel from producing 5G modem chips on time for Apple and contributed to Intel's decision to stop its 5G modem program and sell its entire modem division to Apple (Apple, 2019).

In contrast, the first of Qualcomm's 5G modems and Snapdragon processors became available in 2019; both are fabricated in TSMC's 7-nm foundry (Conway, 2019). It introduced its second-generation 5G modems and processors, including the Snapdragon 865 mobile platform, in early 2020 (Qualcomm, 2020). Qualcomm, a U.S. fabless microchip designer, has been able to regain its leading position in the 5G chip market by relying on TSMC's state-of-the-art chip fabrication capabilities.

Our review indicates that U.S. firms are leading the market in 5G chip designs. However, they must rely on foreign sources in order to have these chips made, as described in the next section. For U.S.-designed chips to be trusted, they must come from trusted foreign sources.

5G Microchip Foundries

Table 4.2 shows the global market share for the largest contract chip manufacturers, or *pure-play fabs*; the capabilities of their most-advanced foundries in process node size; and where they are domiciled.[1] Contract chipmakers make logic chips, not just simpler memory chips (not all memory chipmakers are shown in the table) to the specifications of the chip designer.[2] The most advanced of the foundries shown in Table 4.2 are making the high-value, high-complexity, high-speed 5G microchips listed in Table 4.1: 5G processors and modems. The majority of 5G chip foundries are in Asia.

TABLE 4.2
Major Contract Microchip Foundries

Company	Market Share, as a Percentage	Country	Node Size, in Nanometers
TSMC[a]	54.1	Taiwan	7
Samsung[b]	15.9	South Korea	6
GlobalFoundries[c]	7.7	United States and United Arab Emirates	12
UMC[d]	7.4	Taiwan	14
SMIC[e]	4.5	China	14

SOURCE: Statista, 2020.
NOTES: UMC = United Microelectronics Corporation. SMIC = Semiconductor Manufacturing International Corporation.
[a] TSMC, undated a.
[b] Samsung, 2020.
[c] McGregor, 2019.
[d] UMC, undated.
[e] SMIC, undated.

---

[1] Table 4.2 does not include Intel or its state-of-the-art 10-nm node process foundry because Intel is predominantly an IDM and does not have significant contract chip foundry market share. Some observers claim that Intel's 10-nm process is equivalent to TSMC and Samsung's 7-nm process. However, this argument is immaterial here because Intel has exited the 5G modem market.

[2] Contract chipmakers do not design chips. They only manufacture them, and they do not sell chips directly to device or system vendors, so they do not compete with chip designers—only with other contract chipmakers.

Taiwan Semiconductor Manufacturing Company

The market leader, TSMC, manufactures chips for Apple, Qualcomm, HiSilicon (i.e., Huawei), Mediatek, and other 5G chip designers. TSMC has become a critical link in the supply chain of many 5G companies. Apple, Qualcomm, HiSilicon, and ZTE all use TSMC's 7-nm foundry to produce their 5G chip designs. And as discussed already, when Intel dropped out of the 5G microchip race in 2020, it rendered Apple dependent on Asian foundries for its 5G chips. TSMC was the first company to enter full-scale production of 7-nm chips in 2019 (TSMC, undated c). In terms of revenue, it is the largest pure-play chip manufacturer in the world. TSMC has recently announced that it is on track to begin producing microchips with its new 5-nm process in 2022 (TSMC, undated b).

Samsung

Samsung is one of the leaders in microchip manufacturing. Its V1 semiconductor production line is one of the first to use extreme ultraviolet (EUV) lithography. It can produce chips using its new EUV process node at 7 nm, and it will shortly begin production of 6-nm chips (Samsung, 2020). Samsung used to produce mobile phone processor chips for Apple, but Apple shifted this task to TSMC, fearing that Samsung would misuse Apple's IP to improve its own mobile phone designs. Apple and Samsung have litigated IP issues for years. Over time, Apple has been able to remove Samsung from its supply chain for mobile phones for modems and processors. However, Apple might still be dependent on Samsung for some of its most-advanced display screens, which require unique semiconductor manufacturing processes to make. See Cain, 2020.[3]

United Microelectronics Corporation

UMC is a Taiwanese company that recently built a semiconductor foundry in China and can currently make only 14-nm chips. UMC is an example of a firm that has become untrusted by the United States and by many U.S. firms because of its past behavior. UMC has been indicted for the theft of IP from a U.S. company and for economic espionage. The Federal Bureau of Investigation has indicted a state-owned Chinese company (UMC) and three Taiwanese individuals for an alleged scheme to steal trade secrets from Micron Technology, a U.S. memory chipmaker (Office of Public Affairs, 2018). This case also illustrates the danger of transferring chip designs to a contract chip manufacturer with ties to China.

GlobalFoundries

GlobalFoundries, a U.S.-based company, is owned by a United Arab Emirates holding company. It was formed when IBM decided to sell its leading-edge U.S.-based foundry to outside investors. The original IBM foundry, in Fishkill, New York, was and still is the most advanced foundry in the DoD Trusted Foundry program. GlobalFoundries continues to use this foundry to produce microchips for DoD. GlobalFoundries was once on the leading edge of microchip manufacturing, with chip foundries in Singapore, the United States, and Germany, but it has since dropped out of the race to develop a 7-nm node process foundry, focusing on continuing to produce chips at 12 nm (Shilov and Cutress, 2018). As a consequence, it not a player in the supply chain for advanced 5G microchips—another worrying sign of the loss of state-of-the-art chip manufacturing capabilities in the United States.

Semiconductor Manufacturing International Corporation

The last contract microchip manufacturer shown in Table 4.2 is SMIC, a Chinese company. As the table indicates, SMIC can manufacture chips with a minimum line of 14 nm, just like UMC, the Taiwanese company

---

[3] Apple has invested in Japanese display manufacturers to reduce its dependence on Samsung, but it is not clear that Japanese suppliers are able to produce display screens of the quality and performance provided by Samsung. Chinese companies have moved into the display screen market as well.

with close ties to Chinese state–owned enterprises. SMIC is assessed to be years behind the leading foundry companies (Samsung and TSMC) (Kharpal, 2019).

Intel

Table 4.2 does not include one of the largest U.S. microchip manufacturers, Intel, because it has, until recently, been an IDM—that is, it designs and manufacturers its own chips. Until recently, Intel was not a major player in the contract chip manufacturing market, in which foundries make chips designed by others. However, Intel recently announced plans to enter the contract chip manufacturing market and to make a variety of chips for other companies (Nellis, 2021). This change in strategy was made by its new chief executive officer, Pat Gelsinger, in response to Intel's recent manufacturing struggles. If Intel were on the list, it would be the third-largest microchip manufacturer in terms of revenue, now that it has, in 2021, started producing chips in quantity using its long-delayed 10-nm process. It has announced that it remains committed to producing chips using a new 7-nm process, but this capability might not be perfected for several years; Intel has not yet announced when its first 7-nm chips will be available. Intel has announced that, while it works to catch up to TSMC and Samsung, it will have TSMC manufacture its next-generation, high-performance microprocessors, as an interim measure, to retain its share in this market that is crucial for Intel.

U.S. Microchip Policy

Samsung and TSMC play an increasingly critical role in the 5G ecosystem as the other foundries shown in Table 4.2 fall behind these two leaders. DoD has been concerned about ensuring that it has a source for trusted and unaltered microchips for the U.S. military. At one time, DoD built its own microchip foundry but soon realized that this approach was too expensive. DoD needs few microchips in comparison to the total microchip market, and these small quantities cannot justify the up-front cost of building a state-of-the-art foundry for just DoD systems, which today would likely cost more $10 billion to build.[4] DoD established the Trusted Foundry program in 2005, which began with a single IBM foundry in Fishkill, New York. However, IBM reportedly lost money each year on the foundry and, in 2014, decided to sell the foundry to Global-Foundries.[5] With IBM's assistance, GlobalFoundries remained at or near the state of the art in microchip manufacturing until 2018, when it suspended development of its 7-nm process node. This was effectively an acknowledgment that it was unlikely to catch up to Samsung and TSMC. Nor did its owners have an appetite to fund continued development of the 7-nm process. Consequently, the DoD Trusted Foundry program has remained at 14 nm. Meanwhile, in 2021, both TSMC and Samsung have 7-nm process nodes, and TSMC plans to produce microchips at 5 nm in 2022.

The Trump administration recognized that U.S.-based microchip manufacturers were falling seriously behind their competitors in Asia. In 2020, the U.S. government reached agreement with TSMC to build a $12 billion microchip 5-nm process foundry in Arizona, with production targeted to start in 2024 (Davis, O'Keeffe, and Fitch, 2020).[6] The United States and the state of Texas also have initiated talks with Samsung to expand its Texas operations and to build an advanced microchip foundry in Texas. However, a definitive agreement with Samsung to build a new state-of-the-art fabrication facility in the United States has yet to be reached.

---

[4] DoD demand for chips is so small relative to the commercial market that it would lose substantial sums of money if it were to try to build a state-of-the-art foundry dedicated to its own needs. Its market is now too small for this approach to work.

[5] For a more detailed history of the Trusted Foundry program, see Brackup, Harting, and Gonzales, in production.

[6] By 2024, the 3-nm node will be considered leading edge, leaving the Arizona facility one generation behind but still profitable. See Armasu, 2019.

The Biden administration has recognized the importance of advanced microchips for the U.S. economy and economic competitiveness, as well as for DoD. During the COVID-19 pandemic, the global auto industry has experienced a significant shortage of microcontroller chips, and major automakers around the world have had to temporarily shut down their production lines as a result. The pandemic chip shortage illustrates the risks of depending on chips made by foreign sources, which could be shut down, damaged, or destroyed in a conflict or could become inaccessible because of trade tensions. As a consequence, the Biden administration has proposed subsidizing the construction of advanced microchip fabrication facilities in the United States. The analysis in this report highlights the importance of this initiative, not only to ensure DoD access to state-of-the-art microchips but also to ensure such access to important U.S. technology companies, such as AMD, Amazon, Apple, Google, Microsoft, and NVIDIA.

Chinese Microchip Policy

In its strategic plans, the Chinese government has identified semiconductor manufacturing as a key technology area, one that is important for economic development. As of 2020, SMIC, one of the largest China's largest chipmakers, was assessed to be years behind competitors, but, with the support of the Chinese government, it is not inconceivable that Chinese chipmakers can close the gap with TSMC and Samsung in the next decade. Given the extensive history of Chinese government–sponsored hacking into U.S. and other foreign high-technology companies, it would not be surprising if China were to attempt to steal IP from Samsung or TSMC, purchase start-up companies with important new chipmaking technologies, or hire away science and engineering expertise from Western chipmakers to advance the state of the art of Chinese chip foundries.

Mobile Phone Providers and Operating System Providers

One of the most important parts of the 5G architecture is mobile phones or UE. Mobile phones must be small and lightweight while still providing high-capacity data communications in multiple frequency bands. They must comply with the wide range of frequency allocations of wireless carriers throughout the world. In the 5G era, mobile phones will operate over the widest frequency range, from so-called sub–6 GHz bands to new millimeter wave frequencies at 22 GHz and higher. These features impose challenging technical requirements for mobile phone makers.

Mobile phone capabilities have grown significantly in the transition from 3G to 4G and now to 5G. Each transition has caused consolidation in the mobile phone market. Firms that once were major market players, such as Nokia, Blackberry, and Sony, no longer make mobile phones or have lost significant market share. Tables 4.3 and 4.4 show that, despite considerable changes, U.S. firms continue to play important roles in the mobile phone market. Apple is third in the market by number of units sold. Google is not a major player in the hardware market but supplies the OS used by most smartphones sold today. The dominance of the Google Android OS for mobile phones supports its dominant position in the digital advertising market.

Samsung has the largest market share and has introduced several 5G phones. Samsung is vertically integrated and not only designs but also produces many of its own chipsets for mobile phones.

Apple has relied on Intel and Qualcomm for its modems but designs its own application processors. Apple introduced its first 5G phone in October 2020. Apple might own fewer of its supply chain vendors than either Samsung or Huawei, but it is more profitable. It has relied on microchip vendors that provide best-in-class capabilities in specific areas, such as modems, RFFEs, and audio processing chips. Many other chipmakers that Apple uses are smaller companies, such as Skyworks and Qorvo.

After Intel dropped out of the mobile phone modem business, Apple bought Intel's modem division and might one day design its own modems. This purchase helps Apple ensure that it can maintain access to a secure supply of 5G modem chips in case Qualcomm were to falter or to demand exorbitant royalty fees for

TABLE 4.3

Market Shares, by Number of Units Sold, of Leading Vendors of Smartphone Hardware, August 2021

Vendor	Market Share, as a Percentage
Samsung	18
Xiami	16
Apple	15
OPPO	10
vivo	10

SOURCES: Counterpoint, 2021; Statcounter, undated, queried September 2021.

TABLE 4.4

Market Shares, by Number of Units Sold, of Leading Vendors of Smartphone Operating Systems, July 2021

Vendor	Market Share, as a Percentage
Google	72.2
Apple	26.9
Samsung	0.4
KaiOS	0.4
Unknown	0.1

SOURCES: Counterpoint, 2021; Statcounter, undated, queried November 29, 2021.

its chips. So Apple could also soon become more vertically integrated, like Huawei and Samsung, but, like many others, Apple cannot make its own chips.

Before the United States blacklisted it, Huawei was second in mobile phone market share by number of units sold, and it introduced its first 5G phone before Apple did, in 2019. It attempted to become a fully integrated 5G company. It designs many of its own 5G chips and might own the IP needed to design 5G chips without paying royalties. However, it does not yet have the ability to manufacture its own 5G chips and must rely on TSMC to make the most-advanced chips needed in its leading-edge 5G phones. Since the blacklisting, Huawei's market share has fallen significantly, as shown by its absence from Table 4.3.

The rest of the hardware manufacturers shown in the table are Chinese companies. These Chinese companies rely on other vendors, including U.S. chip designers, for the chips they use in their phones. Nikkei reports, for example, that Qualcomm provides modem and processor chips to many Chinese phone makers other than Huawei (Takano, 2019).

Leveraging U.S. Technology in the Security Competition

ZTE Sanction Violations

U.S. policy toward Huawei and ZTE originally had a domestic focus and prohibited the sale of Huawei infrastructure equipment to U.S. wireless carriers. Almost a decade ago, in 2012, the United States suspected and later confirmed that ZTE had acquired and sold U.S. ICT to Iran and North Korea in violation of U.S. sanctions (Freifeld and Auchard, 2018). While these investigations were ongoing, ZTE was permitted to sell mobile phones in the U.S. market.

In April 2018, DOC banned the export of U.S. technology (i.e., U.S.-designed microchips) to ZTE. This ban caused ZTE to "cease all operations" and could have led to its bankruptcy because of its reliance on Qualcomm and other U.S. chips. Subsequently, ZTE agreed to pay a $1 billion fine, change its leadership, and replace its entire board of directors. As part of the plea agreement, ZTE agreed to have a U.S. compliance officer monitor ZTE compliance with the agreement (Office of Public Affairs, 2017). The compliance monitoring team is staffed by U.S. agents who will remain at ZTE for ten years. This settlement allowed ZTE to resume business and purchase U.S. microchips essential for its smartphones and cellular network products (Davis, Strumpf, and Wei, 2018). However, in 2018, the monitor found that ZTE was not in compliance with the plea agreement. At that time, the U.S. semiconductor export ban was reinstated. However, when the Trump administration reached an agreement with China on a new trade deal, the export ban was rescinded (Morgenson and Winter, 2020).

Huawei Sanction Violations and Intellectual Property Theft Allegations

At about the same time ZTE was charged with violating U.S. sanctions, Huawei was also investigated for violating U.S. sanctions for allegedly selling U.S. ICT to Iran ("Timeline," 2018). Meng Wanzhou, chief financial officer of Huawei and the daughter of its founder, was arrested in Vancouver, Canada, in December 2018. Initially, the United States charged Meng and Huawei with bank and wire fraud in violation of U.S. sanctions on Iran. After reaching a plea agreement with the U.S. Department of Justice, Meng admitted to violating U.S. sanctions against Iran and was released by Canada to fly home to China. She avoided extradition to the United States after the U.S. government dropped these charges (Kharpal, 2020a). Later, the United States levied additional charges on Huawei and Meng. The U.S. Department of Justice indicted Huawei and Meng for racketeering and conspiring to steal trade secrets, technology, and IP from U.S. companies (*United States v. Huawei Technologies*, superseding indictment, 2020). Furthermore, in the indictment, it was revealed that it was internal Huawei company policy to encourage and reward employees to steal IP from competitors or suppliers. In other words, all Huawei employees were encouraged to conduct industrial espionage throughout the world. Because Huawei uses a significant number of U.S. chips, one can therefore expect that Huawei employees would try to steal U.S. technologies and IP while dealing with these companies.

In contrast, ZTE was never charged with stealing U.S. technology or IP. Also, in contrast to ZTE, Huawei has denied all allegations and has not replaced any corporate executives.

Shortly after Meng's arrest in Canada, DoD ordered stores on U.S. military bases to stop selling smartphones made by Huawei and ZTE. In February 2020, U.S. intelligence community officials testified before Congress that Huawei and ZTE smartphones posed a security threat to U.S. customers (Kharpal, 2020a).

In May 2020, the United States expanded the scope of U.S. export controls to require export licenses for the sale of microchips to Huawei, even if these semiconductors were made abroad with U.S. technology (Shepardson, Freifeld, and Alper, 2020). This rule prevents U.S. chip designers, such as Qualcomm, from selling their chips to Huawei without a license. This policy attempts to cut off Huawei from U.S. critical technologies, such as Qualcomm 5G modem chips. The policy also prohibits Google from providing Huawei with its full suite of Android OS components and access to the Google Android app store. And perhaps most importantly, the policy also prohibits foreign chipmakers that use U.S. chipmaking equipment to obtain a license from DOC before they can produce chips for Huawei. This order applies to TSMC and applies to all new orders received by TSMC at least 120 days after May 15, 2020 (Kharpal, 2020b).

In response, Huawei has demonstrated that it can build a 4G phone without any U.S. microchips and without Android. Teardowns of the Huawei Mate 30 4G phone show that Huawei built a 4G phone without any U.S. chips (Byford, 2019). This Huawei 4G phone runs the Huawei HarmonyOS. This Huawei 4G phone contains processor and modem chips designed by HiSilicon and made by Taiwanese and Japanese manufacturers. Huawei lost access to Taiwanese and Japanese foundries in 2020. Because current Chinese chipmakers do not have the capabilities of Taiwanese and Japanese foundries, Huawei might not be able to make 4G phones when its current inventory of Western chips is exhausted.

Furthermore, Huawei has been unable to build a 5G phone without using some U.S. chips ("Huawei Mate 30 Pro Teardown," 2019). One of the last Huawei 5G phones sold, the Mate 20, contained modem chips designed by U.S. and Taiwanese chip designers. Huawei is reportedly dependent on Intel and Xilinx chips for its 5G base stations (Bosnjak, 2020). In recognition of the threat to its business interests, Huawei announced in 2020 that it had purchased more than a two-year supply of U.S. chips to ensure that it could produce 5G base stations and smartphones for at least two years (Li and Ting-Fang, 2020). However, U.S. sanctions against Huawei have had a severe impact on Huawei's mobile phone business. The latest mobile phone announced by Huawei, the P50, does not include a 5G modem and is not capable of operating on 5G networks. It also is not equipped with Android and instead uses the Huawei HarmonyOS (Ting-Fang and Li, 2021).

In contrast, ZTE can still buy U.S.-designed chips by Qualcomm and microchips made by TSMC.

Summary

The United States must continue to account for dynamic change in the relationships and participants of the 5G supply chain to ensure the security of 5G end products, 5G network infrastructure, and mobile devices. As described earlier in this chapter, the 5G supply chain consists of the 5G infrastructure, microchip designers and manufacturers, and mobile phone makers (including the OS). In some segments of the 5G supply chain, the United States has a significant role and presence (e.g., microchip design, OS, mobile phone maker) that can ensure the security of key components. In other areas (infrastructure and microchip manufacturing), the United States is dependent on non-Chinese foreign companies. The United States has a strong position in some parts of the supply chain, where it does not depend on any foreign company, and has the weakest position where it remains entirely reliant on foreign companies in microchip manufacturing. These weaker positions could become even more vulnerable if the non-Chinese foreign companies on which the United States relies fail to compete effectively with Chinese companies in these market segments. The United States should continue to monitor key parts of the 5G supply chain to ensure that the United States maintains access to trusted foreign microchip foundries, and it should establish state-of-the-art microchip foundries inside the United States. Doing so will ensure that the country it is not reliant on untrustworthy microchips for any part of the 5G supply chain.

CHAPTER FIVE

Potential 5G Architecture Vulnerabilities

5G promises technological advancements and opportunities; however, it also invites new vulnerabilities in hardware and software. This chapter characterizes vulnerabilities that could be present in the 5G architecture and examines their potential consequences. In this chapter, we consider vulnerabilities found by investigators, as well as hypothetical potential vulnerabilities a nefarious actor could exploit. The ways in which vulnerabilities could be inserted and exploited in 5G systems highlight the necessity of using trusted partners to develop and operate 5G networks. We show that risks are high if telecommunication hardware or software from an untrusted vendor is used.[1] In this chapter, we address how specific vulnerabilities can be mitigated. This chapter proceeds in four parts and addresses four major areas of the 5G network: the 5G core network, the RAN, 5G mobile phones, and chipsets.

Cybersecurity is a particular area of concern for 5G. Cyber vulnerabilities can occur in the three major segments of the 5G architecture—the core network, the RAN, and in UE or mobile phones—even when a 5G cellular network is connected with one or more other independent telecommunication networks. Such vulnerabilities can be intentionally designed into the architecture or inserted accidentally. 5G components will be software intensive with large software code bases and so could contain a significant number of accidental vulnerabilities or coding errors. These vulnerabilities will likely be discovered over time by the vendor itself or by outside parties. Once a vulnerability is found, it needs to be patched as soon as possible by the 5G system vendor to prevent compromise of carrier networks, customer phones, IoT devices, factory control systems, and, in the future, 5G-connected vehicles.

A trusted vendor would be expected to fix a vulnerability as soon as it is discovered. 5G vendors would also be expected to use cybersecurity best practices in developing software code and integrating hardware components together. In contrast, an untrusted vendor might not fix a software vulnerability immediately, might not use software coding best practices, or might not adopt secure hardware integration approaches. It might instead exploit vulnerabilities as it becomes aware of them to further its own business interests or to support the foreign policy or industrial espionage objectives of its own national authorities. In addition, an untrusted vendor could design in and implement back doors (either permanent or temporary) into its systems to permit unauthorized surveillance or system control. It is this latter set of capabilities that makes an untrusted 5G infrastructure provider so dangerous from a cybersecurity perspective. The 5G infrastructure provider will wield an incredible amount of control and insight over the security features and the strengths and weaknesses of the 5G architecture, even after the system is installed and managed by the wireless network service provider.

---

[1] We define *untrusted vendor* as a company that intentionally implants vulnerabilities, malicious code, or malicious circuitry into its products or that exploits vulnerabilities found in its products.

Core Network Vulnerabilities

The 5G core network is software based and can be instantiated on commercial cloud computing infrastructure. Servers run on virtual machines and are scalable based on network demand. Figure 5.1 illustrates the major elements of the 5G architecture and key 5G core network functions. 5G core functions can reside on multiple servers. In this discussion, we do not examine all aspects of the 5G core network but instead focus on a few key areas that are especially important for network security.

5G networks will provide services to a much wider array of user devices and machines than previous-generation networks have. This includes connected vehicles, and perhaps autonomous vehicles that might rely on the 5G network for ultralow-latency messaging to prevent traffic accidents and ensure high reliability in navigation. The 5G network is also being designed to support IoT communications, which could include building and data center environmental controls, surveillance systems, and factory control systems.

The AUSF network function shown in Figure 5.1 authenticates UE and these other devices to the network. The AUSF verifies the identity of the UE device and the network services available to the user when the UE is authenticated to its home network account (not shown in the figure). This authentication is forwarded to the local network from which the UE is requesting services.

The AUSF handles other security tasks as well, including network slicing security and enhanced international mobile subscriber identity privacy (Palagummi et al., 2018). 5G device and user account authentication is more complex than in the 4G architecture because a wider array of use cases and devices must be supported, as indicated in Figure 5.1. The 5G core network must be able to authenticate IoT devices that might not have subscriber identity module (SIM) cards but instead might use certificates or preshared tokens or keys (Ericsson, 2021). For this reason, the 5G core network employs three types of encryption. 4G networks required physical SIM cards for credentials, so most low-cost IoT devices could not be securely connected to the network. 5G remedies this shortcoming (CableLabs, 2019).

The access and mobility management function, also shown in Figure 5.1, receives all connection- and session-related information from UE and handles connection and mobility management tasks, including the handover of UE from one gNodeB or base station to another (these are shown in Figure 5.1 as antenna towers).

FIGURE 5.1
5G Architecture with Core Network Servers

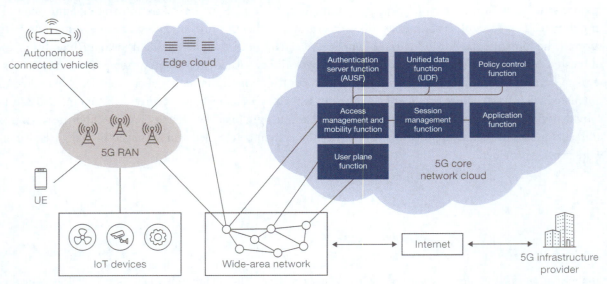

The UDF, shown in Figure 5.1, manages all data associated with a UE or IoT device and the communication session. The UDF will also store all of the encryption keys associated with each UE communication session. In contrast with 4G networks, a 5G network will not only encrypt user data traffic; it will also encrypt all signaling information used to control the user data channel.[2] All the data needed to secure user data and control channels are stored by the UDF. Access to UDF data is controlled by other aspects of the 5G architecture. Other network functions must present a valid data request for user data to the UDF; otherwise, the request will be rejected.

Communications between the network functions shown in Figure 5.1 are encrypted using a separate mechanism to ensure that the 5G core network remains isolated from other processes and workloads running in the cloud.

This short review of the 5G core network reveals that the 5G cyberattack surface is significantly larger than that of a 4G network because of the virtualization of network functions and services and their placement in multitenant computing clouds.[3] In addition, the entire 5G architecture is scalable in the cloud, using SDNs. This scalability provides a cost advantage to wireless carriers but introduces complexity associated with SDNs, as well as the potential for additional software vulnerabilities. Furthermore, the wider array of use cases and devices to which the network will connect necessitates the use of multiple methods for device authentication and encryption to protect user and device identity and message data. These factors increase—from 4G to 5G—the complexity of the security architecture considerably.

Cyberattacks from Untrusted Infrastructure Providers

Figure 5.1 shows that a 5G infrastructure provider could maintain a connection to key nodes or functions in the 5G network for maintenance purposes. Such connectivity has already been implemented in 4G networks and can lead to serious security vulnerabilities. A confidential report by the United Kingdom (UK) consultancy Capgemini states that the 4G wireless network of the Netherlands' carrier, KPN, had been completely compromised by Huawei (at the time, KPN used Huawei equipment in its core network). The Capgemini report "claimed that Huawei could have allegedly been monitoring calls of KPN's 6.5 million users without its [the wireless carrier's] knowledge" (Jowitt, 2021a). Whether this monitoring capability was enabled by known vulnerabilities or by design has not been revealed, but a similar vulnerability was allegedly also found in a Belgian wireless network that used Huawei equipment, which suggests that a systematic capability was engineered by Huawei into its products (Jowitt, 2021a). Given the complexity of 4G and 5G networks, speculation informed by events in the Netherlands and Belgium suggests that Huawei uses its infrastructure connections not only to monitor the health of 4G and 5G networks it supplies but also to conduct surveillance for the Chinese government.

As mentioned earlier in this report, this sort of access is called a *5G network front door*. That is, this form of access has a legitimate business and security purpose but could be misused by an untrusted 5G infrastructure provider using such mechanisms as software updates to insert malware or otherwise conduct espionage and exfiltration operations. Furthermore, key parts of the 5G core network (the UDF) contain encryption and authentication data that could enable the infrastructure provider to conduct malicious activities. These might include targeted surveillance of specific users or efforts to obtain the identity of users that possess specific UE. It is not known whether a cyber back door or vulnerability is needed for an untrusted infrastructure

---

[2] This eliminates some security weaknesses in the 4G architecture that enable malicious actors to precisely geolocate 4G network users using commercial wireless surveillance equipment.

[3] See, for example, Verizon, undated.

provider to gain access to the 5G UDF servers, whether a system debug mode must be set, or whether some type of cooperation is needed to obtain such access.

Weaknesses in the 5G core network architecture might exist but be known by only the 5G infrastructure provider. This knowledge of network vulnerabilities could enable an untrusted infrastructure provider to gain access to the UDF. In addition, the access of the untrusted infrastructure provider to the 5G network could make it possible for the provider to insert malicious code into the core functions of the 5G network without the knowledge of the local wireless carrier.

What could an untrusted infrastructure provider do using the access methods described above? These methods could enable a variety of possible cyberattacks against the 5G network itself or users of the 5G network. Using malicious code or malware, the untrusted infrastructure provider could command a failure in all or a portion of the RAN. This sort of cyberattack would probably be possible to carry out only once and would likely be used only in a war or international crisis. A discovery that an untrusted infrastructure provider had disabled a 5G network of a U.S. or European carrier during war or a crisis would cause significant damage to the brand and reputation of an infrastructure provider, such as Huawei, and would likely cause long-term damage to its business. For this reason, we consider such a cyberattack to be unlikely but nevertheless possible.

A likelier cyberattack by an untrusted 5G infrastructure provider, such as Huawei, would be one designed to conduct espionage, collect confidential user data, or steal IP or trade secrets. As noted in Chapter One, Huawei has been implicated in cyberattacks that had all these objectives.

As noted, an untrusted infrastructure provider could compromise the security of the 5G network and direct user identity data and relevant encryption keys from the 5G network UDM to China. This would enable Chinese authorities to decrypt user message traffic and to track the movements of specific users, such as government officials, Chinese exiles, or Western scientists and engineers. Thus, permitting the use of Huawei equipment in the 5G network core of U.S. wireless carriers or the networks operated by wireless carriers in allied countries would appear to be very risky.

The only methods a U.S. carrier could pursue to attempt to secure its 5G core networks would be constant and thorough monitoring of 5G internal core network functions to ensure that no user or network channel control traffic is diverted to China or intermediate command-and-control servers of Chinese hackers. However, such monitoring could be exceptionally difficult to do because nearly all this traffic is encrypted. Encryption keys would have to be shared with the systems used to monitor 5G network core servers, which, in turn, is a violation of the tenets of the 5G security architecture. Implementing such a monitoring system would probably require a major reengineering of the entire 5G core network architecture. This could be done only in the 3GPP standardization process, and only if led by one of the major 5G infrastructure providers—none of which is a U.S. company. For these reasons, it is unlikely that such an approach would be workable, which, in turn, implies that, if Huawei provided the core network infrastructure of a 5G network, it would be difficult for the host nation or the 5G carrier to prevent Huawei from compromising that network.

Radio Access Network Vulnerability Scenarios

In contrast with that in the preceding section, the discussion in this section is based on hypothetical but plausible scenarios and technical aspects of the 5G architecture. The 5G RAN is based on gNodeB base stations that have a more modular and decentralized design than 4G base stations. In the 5G architecture, some RAN functions are virtualized (not tied to specific pieces of hardware) and can be moved to cloud edge nodes that have much more computing power than earlier 4G base stations have. However, because of this (i.e., virtualization), 5G base station functions (e.g., gNodeB functions) potentially have an increased cyber vulnerability profile.

Some allied countries have considered a policy that permits Chinese companies to provide 5G base stations to their wireless carrier networks and permits only European companies to provide 5G core network components. The question we address here is whether this sort of policy eliminates some of the potential vulnerabilities identified above in the core network and whether this policy is sufficient to secure the 5G network from an untrusted provider of 5G base stations.

In principle, 3GPP technical standards should permit 5G core network servers from one vendor to interoperate with base stations provided by another vendor. A key cybersecurity consideration is whether a base station provided by an untrusted vendor could insert malicious code or exploit a vulnerability in a core network function of the 5G network. The untrusted base station vendor would need detailed technical knowledge of the capabilities of a core network server and the technical design of that core network server. Certainly, a company like Huawei, which produces both base stations and core network components, has such technical knowledge. However, it is less likely that Huawei would have detailed technical knowledge about the software used by a European core network infrastructure provider. Therefore, the likelihood is much lower that an untrusted base station provider could initiate a cyberattack of the kind described in the section above against the core network.

Nevertheless, an untrusted base station provider could still insert malicious code into its own base station products. Such code could be inserted when the devices are manufactured or during software updates. When the attack method is not needed, the undocumented malware could be removed from the system using a software update. When needed, the malicious code could be activated using hidden or nondocumented commands transmitted to the base station from UE (in this case, an untrusted mobile phone, perhaps from the same Chinese manufacturer). This sort of attack could lead to a commanded RAN failure. However, as mentioned above, this sort of cyberattack could likely be conducted only once or twice before it was discovered by the local wireless carrier or by local intelligence services. This type of cyberattack would probably be conducted only during a crisis or war.

Malicious code could also be inserted by an untrusted vendor into its own base station product to collect user data in an unauthorized manner. This type of base station compromise would collect user message traffic for IP theft or espionage against government officials, Chinese dissidents, or others. For this type of cyberattack to be successful, the base station would have to gain access to the encryption keys used to encrypt user message traffic. These encryption keys transit through the base station when the communication session is set up between the UE and the 5G network. They are created by the 5G network core servers. Special code would have to be inserted into the compromised base station to recognize and copy the appropriate encryption keys when the communication session is set up. This would require detailed knowledge of 3GPP technical standards. However, as noted above, Huawei is a major 5G infrastructure provider and has such detailed knowledge. Huawei could develop such a capability and insert it into its base stations. However, this sort of cyberattack would be difficult to develop and require time, money, and skilled programmers. Therefore, we assess that, even if Huawei were to be excluded from the core network of an allied country, if Huawei base stations were permitted in the 5G network, Huawei could still compromise these base stations to conduct espionage and IP theft.

Mobile Phone Vulnerabilities

It is well known that mobile phones can have vulnerabilities. The applications or mobile OS used on the phone can be a source of vulnerabilities that provide a knowledgeable adversary with ways to monitor, record, and disseminate the contents of confidential message traffic to an adversary, to disable the device, or to install malicious applications on the device.

Lithuanian officials have released a cybersecurity report showing that three of the latest 5G phones sold by Chinese mobile phone makers (Huawei, Xiaomi, and OnePlus) have significant software vulnerabilities and cybersecurity issues (National Cyber Security Centre, 2021). The Lithuanian government has asked its constituents to stop using these 5G phones (Jowitt, 2021b). The Xiaomi Mi 9 phone examined has nine vulnerabilities, of which six lead to a risk of personal data leakage. Of the nine vulnerabilities identified for the Xiaomi phone, eight could be exploited by remote communications with the device. Lithuania's National Cyber Security Centre has identified 144 vulnerabilities in Huawei products in the past several years. Twenty-eight vulnerabilities were identified in 2020, and 23 were found in the first half of 2021 (National Cyber Security Centre, 2021). Perhaps the most serious vulnerability found in the Huawei phone is likely present in all Huawei phones sold in Europe. It was found that Huawei's official store in Europe, the AppGallery, sometimes directs users to third-party e-shops in which applications are malicious or virus-infected (National Cyber Security Centre, 2021).

Mobile Phone Operating Systems

In 2020, the United States had a relative advantage in the protection of mobile phones because U.S. high-technology companies developed and controlled the code bases for the OSs used in the majority of mobile phones used worldwide. As noted, Apple has a sizable market share in mobile phones and retains tight control over the OS, iOS, used in its devices. The majority of other mobile phones in use globally, including in China, rely on the Google Android OS, which is also, to a lesser degree, under the control of Google. Both Google and Apple review applications submitted to their app stores for cyber vulnerabilities. Only applications available from the Apple App Store can be loaded on Apple phones and tablets, unless the devices are "jail broken." If an iPhone is jail broken, the user can load apps from third parties on to such iPhones. Google does permit apps to be loaded on to Android phones from app stores not under its control. Third-party app stores might not conduct the same level of cybersecurity review that Google or Apple does. As discussed already, it is documented that third-party Android and iPhone app stores and apps can be a source of cyber vulnerabilities.

Huawei has developed and released a mobile device OS that is based on the open-source version of Android. The Huawei OS is called HarmonyOS. Huawei now also has its own app store. If it chose to, Huawei could easily insert malicious software into its HarmonyOS and into apps in its app store. Such vulnerabilities might not matter if only a small number of mobile devices were affected or if Huawei mobile phones were available only in China. However, the vulnerability could have greater significance for Western countries if Huawei mobile devices, with an OS vulnerability, were to gain significant market share in the United States, Europe, or allied Asian countries, such as Japan or Singapore.

Trustworthy Mobile Phone Chipsets

One of the key components of mobile phones are sets of microchips or chipsets used in the phone. As discussed in Chapter Two, if a 5G microchip came from an untrusted source, a chip in the device could contain malicious circuitry that could be used in a cyberattack. It is possible to insert malicious functionality into a mobile device using malicious circuitry embedded in a compromised chip. As long as the key chipsets used in U.S. mobile phones are designed by U.S. chip designers, the probability that they contain malicious software or circuitry is relatively low.

Mobile Phone Software Malware

Some chipsets contain firmware or software that is loaded into the device or the chip during manufacturing. During the manufacturing process, malicious code could be loaded into a microchip from either a trusted or untrusted source.

A phone manufacturer or designer can scan mobile phones to identify malicious software after the manufacturing process is completed. Although such scanning is unlikely to be exhaustive, it can serve as a deterrent and as a way to detect the presence of counterfeit or defective chips received from suppliers.

A recent case illustrates how malware can be loaded into mobile devices during the manufacturing process. It was discovered that low-cost mobile phones made in China and sold under the Federal Communications Commission (FCC)–sponsored Lifeline program contained malicious software. This software was loaded onto the phones when they were manufactured in China (Brewster, 2020). The phones were sold to low-income American consumers at low prices using subsidies provided by the FCC. In this case, it appears that the purpose of the malicious software was not to conduct espionage but simply to steal identity information and perhaps credit card numbers. Also in this case, it appears that the neither the FCC nor the U.S. wireless carriers that sold these phones scanned the phones for malware. Malware scanning can be done cost-effectively, but ideally should be done at the manufacturing site prior to product packaging. It would be costliest for carriers to do this because the product packages would probably have to be opened and then resealed.

Vulnerabilities in Chipsets

The potential for a mobile phone to contain malicious software or circuitry increases if the chipsets used in the phone are foreign designed and made, especially if the chipset designer is untrusted, as would be the case with HiSilicon, the Huawei subsidiary.

In contrast, U.S.-designed chipsets are unlikely to contain malicious software or circuitry because they are manufactured by foreign foundries that have gained the trust of U.S. companies, both for the quality of their products and for their ability to protect their IP from foreign industrial competitors. The majority of U.S. chip designers are fabless, so they have to rely on foreign manufacturers or foundries, especially for a state-of-the-art chips. As mentioned above, current state-of-the-art foundries exist only in South Korea or Taiwan. These foundries also manufacture chips for Chinese mobile phone makers. So far, these foundries appear to be able to protect the IP of U.S. chip designers and the IP associated with the chip foundry process nodes. However, as mentioned already, there is one outstanding exception to this conclusion: The theft of U.S. IP by Chinese nationals from Micron Technologies in Taiwan serves to illustrate that there is a risk to using chip foundries in foreign countries where Chinese agents and competitors have access. This theft did not result in the compromise of U.S. chips but likely did result in advancing the chip manufacturing capabilities of foundries in China.

The most-damaging chip foundry IP theft would involve technology secrets held by the chip foundry itself: how the foundry configures its chip fabrication equipment and stages its production processes to produce transistors and circuits with the small dimensions needed for state-of-the-art 5G microchips. For these chips, a 7-nm process node is needed. The best Chinese chip foundries today are limited to a 14-nm process node. Technology experts estimate that leading Chinese chipmakers will take a decade to catch up with Western market leaders (Diwakar, 2021). Acquisition of the right kind of foundry IP could shorten this development timeline considerably.

Compromised chipsets can make their way not only into 5G mobile phones but also into 5G infrastructure components, such as base stations and servers. Compromised base stations could be used to support a variety of cyberattacks, as described already. Therefore, U.S. authorities' concerns should extend beyond preventing the use of Chinese base station equipment in U.S. carrier networks and in the networks of allied nations. They also should be concerned about the chipsets used by European 5G infrastructure providers because of the risk that these 5G infrastructure products could contain compromised chipsets from untrusted chipset designers, such as HiSilicon.

This overview of the potential vulnerabilities present in all aspects of the 5G architecture stresses the importance of developing a strategy that ensures that the United States has access to trusted partners needed to develop its 5G network. From 4G to 5G, a new host of vulnerabilities now exist, given the software involved and potential for front doors (and back doors) to be installed by seemingly innocuous vendors. Chapter Six addresses the vulnerabilities highlighted and characterized in this chapter and provides a set of strategic options intended to mitigate the security concerns inherent to 5G. Most importantly, these options focus on how the United States can posture itself and support other trusted vendors to ensure that it does not become reliant on an untrusted vendor for any aspect of its 5G capability.

CHAPTER SIX

Options for a U.S. 5G Security Strategy

The preceding chapters highlight that, in comparison with earlier wireless generations, 5G introduces new security risks that the United States has not previously needed to address—the SDN increased the attack surface of 5G (see the beginning of Chapter Five). As a result, the United States should develop a new approach to 5G that incorporates enhanced security measures to safeguard U.S. 5G technology and capabilities. To do so, the U.S. government will undoubtedly have to play a larger role than it has in past wireless generations. In this chapter, we discuss previous U.S. approaches to wireless generations, explore other countries' approaches to 5G, and then assess each to propose how the United States can develop and implement a strategy for securing U.S. 5G technology.

The supply chain and vulnerability analysis in Chapter Three demonstrates that, in the transition from 4G to 5G, a consolidation of telecommunication providers has introduced new risks and vulnerabilities the United States should be postured to mitigate. The microchip market has experienced a significant consolidation of suppliers able to manufacture the state-of-the-art chips needed for 5G networks. And the potential exists for a similar consolidation in mobile phone suppliers as well, if chipsets needed for 5G phones were to be withheld or priced exorbitantly by chipset designers or makers. Further, the trusted foreign companies on which the United States must rely for much of its 5G infrastructure might not be able to compete with Huawei in the future. Foreign infrastructure providers, such as Ericsson, Nokia, and Samsung, could lose market share and fail, or they could fall victim to consolidation. There is ample evidence that Huawei's use of state subsidies has given it an unfair competitive advantage in the marketplace.

As noted above, Huawei's growth within the market poses risks beyond just subsuming trusted vendors, such as Ericsson and Nokia. Huawei does not provide a viable option for the United States because of its history of espionage and IP theft and its apparent unwillingness to change. Additionally, key issues of chip foundries must be accounted for in the U.S. deployment of a 5G capability. A U.S. security strategy will involve more than speed—it must secure the long-term integrity of the 5G supply chain. Securing the U.S. 5G supply chain will require additional funds and will require new investment to support R&D in cellular communications and associated cybersecurity architectures. The United States must develop a strategy that accounts for these unique 5G security concerns stemming from the supply chain and the need to maintain trusted suppliers for its 5G infrastructure.

The U.S. experience with 4G provides a baseline that highlights several elements that a country should account for in developing and securing its next-generation wireless networks. The 4G approach also suggests what might be new about 5G. This chapter first discusses the U.S. approach to 4G and then introduces several options the United States might pursue to secure its 5G infrastructure and supply chain. These options include an extension of the U.S. strategy for 4G, the current UK approach, and different variations of a fully trusted 5G architecture option. We describe these in greater detail in the next section.

The Past U.S. Approach to 4G

The U.S. approach to 4G consisted of an increase in industry investment coupled with government policies that enabled the flexible use of spectrum (Recon Analytics, 2018). 4G did not present the same set of key actors, nor did the leading actors suggest untrustworthy behavior that would incentivize the government to play a more active role in the development of wireless communication technology. The U.S. government did not provide significant R&D investment to private companies to spur 4G innovation, nor did the government play a key role in determining what vendors would be included or excluded from providing infrastructure components.

The United States also used a free-market approach for 4G, allowing companies to compete and develop 4G infrastructure with minimal government intervention. Some suggest that the United States should extend this approach to 5G; however, others contend that the security vulnerabilities that 5G introduces require a more nationalized strategy (Downes, 2019). Many argue that U.S. industry led in the rollout of 4G thanks to free-market policies that enabled innovation and competition (McGill, 2018). The Cellular Telecommunications Industry Association used the U.S. approach to 4G to argue that the U.S. government should allow the free market to encourage competition and innovation among private companies for 5G. CTIA argued that, because the U.S. government did not intervene in the 4G rollout, it allowed private companies to compete and innovate, resulting in a U.S.-led 4G industry (Gillen, 2019).

In contrast, others have used the security risks of 5G and potential for China to be a leader in 5G infrastructure as evidence that the United States should depart from the 4G approach and instead adopt more-nationalized policies (Crawford, 2018). Many of these arguments for nationalization stress the need for more-rapid innovation and technological development to keep pace with China. Some argue that U.S. companies and wireless carriers continue to move too slowly (Barfield, 2019). These debates over the applicability of a 4G approach for 5G suggest key considerations for the United States in developing a security strategy. We explore these issues in the next section.

The Key Elements of a 5G Strategy

The U.S. experience with 4G and the debate over how the country should subsequently pursue 5G raise questions of the role and degree of government involvement, type of government policies, and treatment of foreign vendors in developing the telecommunication infrastructure. Although it was not apparent in the U.S. 4G rollout, the United States must now also consider the role of allies in developing and implementing a 5G strategy and focus more on securing its supply chain.

5G technology, infrastructure, and supporting elements, such as spectrum availability, span across various U.S. government agencies and departments. Any successful strategy to secure 5G will need to account for these different institutions. In Table 6.1, we briefly discuss U.S. government departments and agencies with a role in U.S. 5G.

In developing and implementing a strategy to secure the U.S. 5G infrastructure, the government—which includes all the stakeholders listed in Table 6.1—must account for the following key elements:

- the role of government funding and investment for R&D
- development of secure 3GPP technical standards
- federal policies that underpin the 5G infrastructure (spectrum, base stations, etc.)
- treatment of foreign 5G infrastructure vendors
- supply chain security
- role of and coordination with allies.

TABLE 6.1

Current 5G-Related Authority and Activities of Federal Entities

Organization	Function	5G Authority	5G Activity
BIS	Ensures U.S. export control and treaty compliance system in support of U.S. strategic tech leadership	Entity lists and sanctions	Placed Huawei on entity list in 2019
CISA	Manages cyber and physical risk to U.S. critical infrastructure	Is in charge of security and resilience of 5G technology and infrastructure	CISA 5G strategy
FCC	Regulates communications by radio (including cellular), television, wire, satellite, and cable in the United States	Oversees allocation and sale of wireless spectrum; infrastructure policy; regulations	Midband (C-band) spectrum allocation and auctions for 5G networks
NIST	Advances science and engineering measurement standards and technology	Advisory role	NIST alliance for 5G networks; examination of 5G standards for cyber vulnerabilities; research on 5G spectrum-sharing standards
NTIA	Advises the President on telecommunication and information policy issues	Advisory role	5G implementation plan for *National Strategy to Secure 5G*; research on advanced civilian communication systems
USPTO	Grants U.S. patents and registers trademarks; advises the President, Secretary of Commerce, and federal agencies on IP policy, protection, and enforcement	Can grant (or deny) patents filed for 5G technologies	Patent review

SOURCES: Bureau of Industry and Security (BIS), undated; CISA, 2020; FCC, undated; NIST, 2021; NTIA, 2021; USPTO, undated.

The role of government funding and investment for R&D refers to how the government can assist U.S. private-sector organizations and U.S. SSOs to support 5G technical standards and technologies and how it can assist U.S. academic institutions and companies in developing advanced wireless communication and networking technologies.

Government R&D will also include resources provided to government R&D organizations, such as NIST and ITS to evaluate important properties of 5G technical standards and NSF for basic research in advanced wireless technologies. As discussed, the government played a minimal role in the R&D for 3G and 4G. As a result, NIST, ITS, and NSF research portfolios in wireless communications have historically been small. Given the security considerations that 5G has amplified in telecommunications, the government has demonstrated an increased emphasis on the development and deployment of 5G.

Thus, the second element, federal policies, will also contribute to how the United States develops 5G and at what pace. Many government bodies, such as the FCC, control electromagnetic spectrum in the United States. The policies these bodies implement will dictate when and how 5G can be deployed (spectrum must be made available in the C-band to realize the full benefits of 5G). Spectrum allocation and bandwidth dictate how far apart base stations will need to be spaced, which will influence the cost to wireless carriers for deploying 5G infrastructure.

Third, the U.S. approach to 5G must consider how it treats foreign 5G infrastructure vendors. For example, the United States will need to determine which companies to exclude or include and how it treats foreign vendors that have been included in an allied country's 5G infrastructure. The reputation of foreign vendors and their willingness to provide security-related information on their products will be important considerations. Their willingness to share with U.S. government officials the sources of the chipsets used in their products—both who designed them and where they were manufactured—will be increasingly important.

Foreign 5G infrastructure vendors should be encouraged, but not required, to use U.S.-designed chips. Just as in automobiles, it might be appropriate to not necessarily impose U.S. content rules on 5G infrastructure components manufactured overseas but to set goals that vendors could strive to achieve. This would also enable U.S. officials to track the use of foreign chipsets in U.S. 5G networks.

Fourth, the U.S. approach should also incorporate measures to secure its 5G supply chain. The United States must monitor the security of its 5G supply chain and ensure that major vendors in the supply chain remain trusted. Doing so involves ensuring that vendors do not become victims of cyber or physical attacks that could interrupt the flow of trustworthy 5G systems or components to end users, U.S. wireless carriers, and those in allied countries. As discussed in Chapter Four, U.S. companies in the 5G supply chain depend on third-party vendors in Asia for state-of-the-art microchips necessary for 5G systems. The U.S. government has recognized this dependence and has initiated legislation to support the construction of leading-edge microchip foundries in the United States to ensure trustworthiness.

Last, the United States must consider the role of allies and how they decide to secure their own 5G networks. Allied countries, specifically those where U.S. military installations reside, could affect U.S. communications and either invite or mitigate vulnerabilities.

A U.S. strategy for 5G security must incorporate these elements. As discussed throughout this report, many of these elements have emerged in the transition from 4G to 5G—specifically, concerns about the supply chain and rise of Chinese vendors. The next section outlines several options and explores the merits and downsides of each approach.

Options for a U.S. Approach to 5G

An Extension of Past U.S. Approaches to 3G and 4G

In the transition from 3G to 4G, little changed in the government approach except that the United States imposed an informal ban on Huawei equipment. The informal ban was in response to Huawei stealing IP from Cisco Systems. Although the United States chose to informally ban Huawei, this ban did not apply to all U.S. wireless carriers. Huawei 4G infrastructure exists today in the wireless networks of many small, rural carriers. In addition, the United States did not require or request allies to do the same. In fact, the United States did not have a full ban on Huawei, still allowing some Huawei routers in defense networks and in rural areas not deemed high risk. Certain laws and regulations made it difficult for the United States to pursue a complete ban of Huawei equipment.

An extension of U.S. approaches to earlier-generation cellular networks (e.g., 3G and 4G) rollouts would consist of minimal government involvement in the development of 5G (Nunno, 2003). For this option, the government would likely not contribute significant investment for R&D, nor would it become involved in the markets, except for informally banning Huawei from major parts of the U.S. telecommunication infrastructure. In the transition from 4G to 5G, Chinese vendors have increased their global market share, and the risk of continued consolidation looms. As a result, although the government might want to pursue the same approach that it used for 4G, it has openly acknowledged the need to account for the security implications of Chinese telecommunication companies' continued rise in market share (Dano, 2020).

In this option, the United States would rely on a wide array of foreign vendors for the 5G infrastructure and would let these vendors manage their own supply chains for microchips and other components. As shown in Chapter Two, almost every option would include a U.S. reliance on foreign vendors. The key choice facing the United States is how to treat these foreign vendors and how to ensure that it uses trustworthy telecommunication vendors to construct the U.S. 5G network.[1] The United States does not have a domestic telecommunication company that can supply 5G infrastructure and would thus rely on European or South Korean providers for the majority of its 5G infrastructure. Additionally, as with 4G, this approach would not include requests that U.S. allies also ban Huawei equipment.

This option offers some benefits and important trade-offs. If pursued, this option would require very little investment from the United States. However, this option suggests a litany of security implications. In particular, it could lead to potential security vulnerabilities in allied 5G networks, a risk that would grow over time. In the event that the United States cannot secure Nokia or Ericsson 5G infrastructure, or if those companies fail, what would the United States do? Would the United States rely on Samsung instead? Currently, Samsung has very limited market share in the global 4G and 5G markets. In addition, if the United States were to rely on Samsung for 5G infrastructure, that reliance could provide an unfair boost and competitive advantage to Samsung in the mobile phone market, which could lead to competitive challenges for other important U.S. companies, such as Apple. Furthermore, and perhaps most importantly, the United States would need to trust Samsung as a telecommunication provider. Last, the companies on which the United States relies for 5G, which would be foreign under this option, could have repercussions for other companies important for the U.S. economy, including Qualcomm, Apple, and Skyworks Solutions. A dramatic consolidation in the 5G infrastructure market could have far-reaching effects on related markets in which U.S. companies currently play important roles in the 5G ecosystem and in the U.S. economy.

The Current U.S. Approach to 5G

The current U.S. approach to 5G includes more government involvement than with 4G to secure the supply chain and exclude untrustworthy telecommunication equipment providers. In this option, the U.S. government would continue a formal ban on Huawei equipment, consistently with the National Defense Authorization Act for Fiscal Year 2019 (Pub. L. 115-232, 2018). Further, this option includes a formal request that allies ban Huawei telecommunication equipment in their networks to uphold the same security measures as the United States (Woo and O'Keeffe, 2018).

The current approach is based on the five lines of effort identified in the White House strategy for 5G (White House, 2020). A variety of activities have been identified in the NTIA, and the 5G implementation plan supports these lines of effort (NTIA, 2021). One key activity is to increase R&D in 5G and 6G wireless communications to ensure U.S. leadership in this technology area. Another is to assess threats and vulnerabilities to 5G infrastructure and to identify security gaps in the supply chains of 5G systems. Yet another activity is to identify incentives and policies to close security gaps found in the 5G architecture. And finally, another activity is to promote U.S. leadership in development of international standards for 5G.

The current approach also involves key efforts from the government to secure its 5G supply chain, as shown in Table 6.1. These efforts primarily involve the United States encouraging other companies and countries to prevent Huawei from advancing its technological capabilities and slowing its moves toward market dominance. However, we have seen more-direct actions from the U.S. government through BIS placing Huawei on its entity list and NIST's development of cybersecurity guidance on 5G, although the latter remains in progress. Moreover, recently, the United States has tried to prevent Huawei from contracting with TSMC for the production of its chip designs and to force Huawei to manufacture them inside China, which would be costly and technologically challenging (Kharpal, 2020b). The United States has also formally supported Qualcomm despite past district court rulings that could undermine this support. Recent Federal

---

[1] Diplomacy could be an important avenue for improving the cybersecurity of 5G networks, but determining exactly what sort of diplomatic actions would lead to this goal requires further research.

Trade Commission antitrust rulings might require Qualcomm to share IP with Huawei, benefiting Huawei while hurting Qualcomm (Thompson, 2020). Another key element of this approach is to bring foreign state-of-the-art microchip foundries to the United States. The United States has recently announced an agreement with TSMC to develop a 7-nm foundry in Arizona and have it operational by 2024 (Davis, O'Keeffe, and Fitch, 2020). This agreement will help secure the microchip supply chain of U.S. 5G companies and for U.S. government programs. In this option, the United States would likely pursue additional efforts to secure its supply chain and support trusted vendors, such as Qualcomm.

Although the current approach might prove a suitable option, several issues still remain. First, the United States still needs a concrete and resourced strategy to secure its microchip supply chain.

Second, although DoD and FCC have made progress with the recent auction of C-band spectrum for 5G, the United States has allocated less midband spectrum to 5G than other countries in Europe or Asia have (WG on Trust and Security in 5G Networks, 2021). More midband spectrum is needed.

To achieve a more secure supply chain while also moving swiftly to develop and deploy a U.S. 5G technology, several questions and options emerge for this approach. Does the United States want to use punitive measures against Huawei and continue to pressure TSMC to prevent Huawei from using TSMC chip foundries? More importantly, how can the United States take measures to weaken Huawei economically to limit its technological advance in a strategically prudent fashion? Huawei has recently faced challenges by no longer selling 5G smartphones, which has negatively affected its financial position. In August 2021, Huawei reported that its revenue had dropped 38 percent in the preceding quarter (Strumpf, 2021). Huawei's inability to produce phones that can run on 5G comes as a direct result of U.S. sanctions restricting Huawei's access to certain chip technology (Ting-Fang and Li, 2021).

And what should the U.S. approach be toward other Chinese 5G infrastructure and mobile phone vendors, such as ZTE, Xiaomi, OnePlus, and OPPO? The United States has not accused these companies of stealing IP or subverting U.S. sanctions against Iran, and they could present lower cybersecurity risks to U.S. allies.

A Proposed UK Approach

The UK proposed an alternative approach to 5G in 2019, which presents another option. The UK has chosen to include Huawei in parts of the network and has not developed a formal supply chain policy. The UK was originally going to allow Huawei into its 5G infrastructure but subsequently stated that it would exclude Huawei equipment from entering its core network (Kelion, 2020a). Recent reports note that the UK's prime minister, Boris Johnson, reversed this decision and decided to phase out the use of all Huawei equipment by 2027 (Kelion, 2020b). The UK justified the use of Huawei from an economic, political, and security standpoint. Economically and politically, the UK noted, banning Huawei entirely would hurt UK telecommunication providers and slow the UK rollout of 5G. The government stated its intent to provide high-speed network connectivity, and excluding Huawei would likely slow its ability to fulfill this expectation (Nuttall, 2020). From a security standpoint, the UK suggested, by preventing Huawei from entering the sensitive core network of 5G infrastructure, it mitigates several vulnerabilities that the foreign vendor could introduce. From a cybersecurity risk standpoint, the original UK approach of excluding Huawei only from the core network reduced the risk of Chinese espionage and IP theft significantly but did not eliminate it entirely.

This approach stems from a government effort to minimize the economic costs of developing 5G given the lack of viable telecommunication equipment providers. The UK does not supply any of the technology or systems needed for 5G. In allowing Huawei to provide some equipment for their 5G infrastructure, the government intends to promote competition among existing telecommunication companies and reduce costs. Additionally, the UK recently left the European Union and might not feel a need to support Ericsson and Nokia. Its approach thus far suggests that the government has some confidence in its ability to secure the 5G infrastructure by heavily monitoring Huawei equipment in certain parts of the network. However, the vulnerability analysis presented in Chapter Two suggests that it would be very difficult and costly for the UK to secure its 5G infrastructure even if it permitted the use of Huawei output only for base stations and mobile phones.

This option perhaps provides only an illusion of an economic incentive but introduces significant security concerns. From an economic standpoint, using Huawei could encourage competition, allow a country to deploy 5G more quickly than it could otherwise, and develop the infrastructure at a reduced cost. However, from a security standpoint, using Huawei invites additional security vulnerabilities, given that the UK is not likely to consider Chinese 5G infrastructure providers to be trusted 5G vendors.

Germany might take an approach similar to the one proposed in the UK in 2019, to permit Huawei into some parts of German 5G networks. Germany is under substantial pressure from the Chinese government to allow Chinese 5G infrastructure vendors to participate in building 5G networks in Germany. Chinese diplomats have stated that China would rethink its policy of permitting German car companies to sell cars in China if Chinese companies were not permitted to compete in the German 5G market. Not surprisingly, larger trade issues with China have come into play in 5G security policy debates in allied countries. So far, in the United States, the Huawei ban on providing 5G and other network infrastructure has not been linked with the larger set of trade issues under debate in Washington or Beijing.

A Fully Trusted 5G Architecture

The United States should consider an option that provides a fully trusted 5G architecture, from both offensive and defensive perspectives. From a defensive perspective—meaning that the United States would pursue only those measures that defend its networks rather than actively degrading bad actors' capabilities—the U.S. government would need to implement several policies. To achieve a fully trusted 5G architecture, the government would need to ban Huawei and ZTE equipment from all parts of the network. The United States also needs to work closely with its 5G trusted partners (e.g., Ericsson, Nokia, Samsung) on 5G security issues to provide technology assistance and ensure the integrity of 5G technical standards. The United States should also work with trusted 5G infrastructure providers to ensure the integrity of its 5G microchip supply chains.

The United States could take similar steps in securing mobile phones sold in the United States. It could ask for chipset transparency from 5G mobile phone vendors and mobile phone retailers, including U.S. wireless carriers. It could also encourage the use of U.S.-designed chipsets in mobile phones sold in the U.S. market. It should not exclude chipsets designed or produced from trusted foreign vendors in Japan or South Korea, but it could track the market share of 5G chipset designers and manufacturers to ensure the economic health of key suppliers in the U.S. 5G supply chain. Congress could also pass legislation that enables the FCC or DOC to sanction and fine foreign mobile phone vendors that supply mobile phones to the U.S. market that have malware loaded on them when they are assembled. Foreign manufacturers that supply compromised phones to the U.S. market could be also added to the DOC entity list and banned from all future federal and U.S. commercial procurements. In addition, Congress could pass legislation that requires U.S. mobile phone designers and retailers to attest to the chipset content of their offerings and whether their phones contain any chip manufactured or designed in China.

Supply chain security must also be incorporated into any strategy for U.S. 5G. How the United States chooses to secure the 5G supply chain can vary, from complete bans to informal exclusions to actions that try to actively undermine a foreign vendor. Supply chain security approaches must consider not only the designers of 5G chipsets but also where they are manufactured and whether these manufacturing facilities and contract chipmakers are willing and able to protect the IP of U.S. chip designers. 5G supply chain transparency should be increased for all major components of the 5G architecture: core network virtual servers (software sources), base stations, edge compute nodes, and mobile devices. Manufacturers of 5G hardware should be required to declare to the U.S. government which chipsets they use, where those chipsets were manufactured, and whether their products include any chips of Chinese origin. CISA is developing supply chain best practices through the ICT Supply Chain Risk Management (SCRM) Task Force that 5G infrastructure vendors for U.S. carriers should be required to follow.[2]

The theft of U.S. IP and leakage of U.S. chip design prowess to foreign adversaries can have repercussions that extend beyond the security of 5G infrastructure and could include implications for military microelectronic systems. As noted in Chapter Five, all the state-of-the-art foundries needed to make 5G chips are in Asia. Another potential vulnerability of the U.S. 5G supply chain is the theft of chip manufacturing IP from TSMC and Samsung. The manufacturing secrets and prowess of these companies must be protected from adversaries. The United States can help protect TSMC and Samsung manufacturing IP by incentivizing these companies to locate their new foundries in the United States where they can be better protected by U.S. intelligence and law enforcement agencies.

The steps taken by the current administration to onshore a state-of-the-art TSMC chip foundry on U.S. soil is a major step not only in securing the U.S. 5G supply chain but also in providing a trusted source for secure chips needed in U.S. military, space, and intelligence systems. This step will help secure the IP of U.S. chip designers and the technical lead these designers have over HiSilicon and other Chinese companies. In the past several decades, DoD has tried to ensure a domestic source of secure state-of-the-art chips for military applications, using the DoD Trusted Foundry program. However, this program was not able to keep up with the leading commercial contract chip foundries, such as TSMC, and today, no U.S. chip foundry has the capabilities of Samsung or TSMC. DoD and the rest of the U.S. government have recognized this fact and their reliance on foreign technology. The United States could also offer assistance to the Taiwanese government and TSMC to protect their foundries in Taiwan from foreign adversaries and Chinese competitors.

Even though TSMC will build an advanced foundry in the United States, it will likely continue to conduct its most advanced R&D in Taiwan. The United States should not rely on TSMC alone to ensure that U.S. 5G companies and, in particular, U.S. chip designers can access a supply of state-of-the-art microchips. The recent problems that Intel has experienced in its chip foundries and the implications of Intel's recent delays in perfecting its 7-nm process illustrate the challenges faced by the entire U.S. digital economy and DoD (Alcorn, 2020). Intel's foundry difficulties have, in turn, led to major problems for Nokia, one of two U.S. trusted 5G infrastructure suppliers, and has caused Nokia to search for alternatives to Intel's 5G field-programmable gate array chips (Morris, 2020). To ensure the financial and technical health of trusted 5G suppliers, the United States might need to help restore technology leadership at Intel's foundry division or facilitate the sale or merger of this division to a U.S. company that restores U.S. technology leadership in this increasingly important market.

Summary

We provided a set of options in this chapter for how the United States could approach 5G. In this summary, we evaluate the strengths and weaknesses of each approach and offer our recommended option: a fully trusted 5G architecture.

Previous U.S. Approaches to 3G and 4G

As noted above, previous U.S. approaches to 3G and 4G saw little government involvement in the rollout of earlier generations of the U.S. cellular network. However, with 3G and 4G, the U.S. government did not face

---

[2] For more information on the ICT SCRM Task Force, see CISA, undated.

the same security risks it does with 5G. In addition, with the recent increase of ransomware attacks against critical infrastructure (e.g., the Colonial Pipeline attack) and sophisticated nation-state cyberattacks against U.S. government and corporate networks, there is growing concern about the cybersecurity of U.S. critical infrastructure. As a result, the United States should not step back and adopt its previous approach to 3G and 4G because, at a minimum, it does not allow enough government oversight of the cybersecurity of the network.

The Current U.S. 5G Approach

The current U.S. 5G approach does incorporate security concerns—in particular, by banning Huawei products from the U.S. 5G network. It also bans Huawei from using U.S. technology in its 5G products, in an attempt to slow the rise and possible dominance of Huawei in the 5G market. The current U.S. 5G approach also attempts to secure the supply chain of U.S. 5G companies by encouraging TSMC to build a 7-nm process in the United States. However, important elements are still missing from the current approach. The United States is not doing enough to ensure that it has long-term access to a trusted 5G infrastructure vendor and system integrator. The U.S. government might also not be doing enough to support R&D in 5G and next-generation wireless communications and to ensure that U.S. companies retain technology leadership in key 5G technical areas. Although U.S. companies have technical leadership in several key areas of the 5G supply chain, the United States is dependent on foreign suppliers in other areas, and changes in the supply chain could put leading U.S. companies at a disadvantage over the long term. In particular, if U.S. 5G chip designers (e.g., Qualcomm, Apple) were to lose access to state-of-the-art chip foundries, their position in 5G markets could be threatened.

The UK 5G Approach

The original or the new revised UK approach to ensuring the security of UK 5G networks also has significant downsides. Although the revised approach is a significant improvement over the original one, it still has shortcomings. The revised UK approach does not explicitly account for the financial health of European 5G infrastructure providers. If these providers were to go out of business, the UK could become reliant on untrustworthy infrastructure providers. Although the UK is in a similar situation to that of the United States in that no UK-based companies provide 5G infrastructure products or mobile devices, it still should be concerned about the security of foreign products it needs for its network. Additional steps, outlined in the next section, are needed to ensure the long-term viability of trusted 5G infrastructure products and mobile devices.

A Fully Trusted 5G Architecture Approach

The first element of this approach—a U.S. strategy for securing U.S. and allied-country 5G infrastructure— should include steps to ensure that the IP of Western 5G technology developers is incorporated into 3GPP SEPs, that the IP of U.S. companies is protected during the development of 3GPP technical standards, and that Western companies retain technical leadership in critical areas of wireless communications that enable 3GPP SEPs to be based on U.S. IP.

Second, the United States should not hinder two U.S. companies, Apple and Google, from retaining their leadership positions in the OS market. The United States should monitor the market share of Chinese companies—in particular, Huawei—in the mobile phone market and point out when their products have serious cybersecurity shortcomings. The United States should monitor 5G OS development to ensure that Huawei does not develop a viable OS alternative to Apple iOS and Google Android.

Third, the U.S. government should develop SCRM criteria and supply chain review processes to distinguish between trustworthy and untrustworthy suppliers for 5G infrastructure. The United States should share these criteria and counterfeit- and compromised-component alerts with trusted 5G infrastructure vendors. Companies that sell mobile devices in the U.S. market should attest to the chipset content of their offerings and whether their phones contain any chips manufactured or designed in China.[3]

Fourth, this approach should ensure that U.S. chipmakers are leaders not only in the design of microchips but also in the design of secure logic chips. NSF has funded research into the development of secure microprocessors and other microchips, but adversary techniques for chip compromise continue to advance (NSF, 2014). This R&D will provide a hedge in case U.S. efforts to establish state-of-the-art microchip foundries in the United States falter. One challenge with this R&D objective is to encourage U.S. microchip designers to collaborate with U.S. academics working on trustworthy chip designs and to secure the transfer of the associated IP to U.S. companies.

Fifth, the U.S. government should support R&D into 5G and next-generation wireless technologies and chip design, to ensure that U.S. companies have a strong presence in the 3GPP process of developing technical standards and in 5G microchip markets. Also for this R&D objective, the U.S. government should put in place measures to ensure that associated IP for these inventions is transferred to U.S. companies.

Sixth, because U.S. microchip manufacturers have fallen behind foreign competitors and, at the same time, China has made the development of its microchip industry a national priority for the development of its economy, the United States should carefully monitor the financial and technology health of the few remaining U.S.-based microchip foundry companies that produce logic chips.

Last, we recommend that, in this approach, the U.S. government consider a series of options to create a trusted domestic source of state-of-the-art microchips for 5G, U.S. computer makers, and the U.S. government. By *state of the art*, we mean foundries capable of building chips with 7-nm or smaller transistor features. These new U.S. state-of-the-art foundries might include foundries built by at least two of the following companies, and perhaps more than two if funding permits:

- TSMC
- Samsung
- Intel
- GlobalFoundries (if it becomes a publicly traded company on a U.S. stock exchange).

This list of companies is based on the analysis presented in Chapter Three. These are the only companies in the world today that are or potentially have the technology base to produce logic chips at 7 nm or less, if they can secure the capital needed.

TSMC has already broken ground on a new microchip foundry in Arizona, but the United States should have access to more than one such foundry and, if possible, one that is operated by a U.S. company. This option would use federal and state government incentives to help defray foundry construction costs in the United States, as specified in current draft legislation (i.e., the Creating Helpful Incentives to Produce Semiconductors for America Act, or CHIPS Act).[4]

We believe that only the fully trusted approach can provide all the capabilities needed to ensure that future U.S. 5G networks will be secure. We further believe that only this approach will enable U.S. industry

---

[3] Memory chips could be excluded from this requirement.

[4] As of September 2021, the CHIPS Act had not been passed but, as written, includes $50 billion for microchip R&D, development of advanced microchip manufacturing, and incentives to qualified companies to build new state-of-the-art microchip foundries in the United States.

to retain technology leadership in the key areas to ensure the security of U.S. telecommunication networks in the long term.

CHAPTER SEVEN

Findings and Recommendations

The 5G Security Competition

The competitive landscape in U.S. telecommunications has traditionally been viewed through the lenses of economics and technology. The security of U.S. networks was less of a concern because infrastructure suppliers were U.S. or European and presumed trustworthy. Perhaps it was also assumed that U.S. wireless carriers could make U.S. cellular networks secure by modifying or configuring equipment they bought from suppliers. However, 5G networks are now more complex and have many capabilities, settings, and possible configurations. Much of this complexity is reflected in the microchips that can be found in 5G systems. More than 80 percent of all microchips are now made in Asia, and a growing number are made in China. According to industry projections, by 2030, the United States will make less than 10 percent of the world's microchips, while China and Taiwan together will make more than 40 percent of them (Ip, 2021). Microchips make their way into 5G products in a complex supply chain with many steps and multiple opportunities for microchips to be compromised along the way. Cybersecurity risks might increase significantly when 5G is deployed worldwide unless more-careful attention is paid to the security of the 5G supply chain.

China has used its leading cellular telecommunication company, Huawei, as a means to surreptitiously collect sensitive national security, foreign policy, and IP information around the world. Huawei was caught spying on the African Union, convicted of stealing software code from U.S. companies, indicted by the U.S. Department of Justice for the theft of U.S. company trade secrets, and assessed to be capable of gathering mobile phone user data at scale using its cellular infrastructure equipment deployed in the Netherlands and Belgium. Huawei is subsidized by the Chinese government in many ways, so it can sell its 5G infrastructure products at deep discounts that its Western competitors cannot match. Huawei and ZTE might increase their penetration of the cellular network infrastructure market as 5G is deployed globally, providing China with significant intelligence and illicit economic advantages.

Competition between the United States and China in 5G now involves more than economic markets and technology leadership. Also at stake are the cybersecurity and integrity of next-generation cellular networks. These three dimensions of the 5G competition—economics, technology, and security—are intertwined. The companies that lead in 5G technologies might gain market share, regardless of their security strengths or weaknesses or their trustworthiness, unless cybersecurity and trustworthiness are determined to outweigh technology or economic preferences. Similarly, if a company can sell 5G products at lower prices than competitors can, it will likely gain market share, unless security is made an overriding concern. An untrustworthy company, such as Huawei, can leverage its market position and technology to help China achieve its national security goals and compromise the security of other nations. The security advantage that China might be able to derive from 5G networks will rely on the market position and technological capabilities of its 5G companies.

Table 7.1 summarizes our analysis of the market, technology, and security positions of Chinese and U.S. companies in 5G, including their access to the 5G supply chain. The table indicates where Chinese or U.S. 5G companies have a market or technology advantage (designated by A_M or A_T, respectively) or a market or

TABLE 7.1

Status of the 5G Security Competition Between China and the United States

5G Architecture	China	United States
Network infrastructure	A_M	D_M
Mobile devices	A_M	A_T
Mobile device OSs	D_T	A_M, A_T
Microchip design	D_T	A_M, A_T
Microchip foundries	D_T	YD_T

NOTE: Dark green indicates that the country has a strong advantage. Light green indicates that the country has a slight advantage. Yellow indicates that the country relies on foreign third-party suppliers because it lags in the technology area but has access to products made by foreign suppliers.

technology disadvantage (designated by an D_M or D_T, respectively). The 5G architecture areas we considered are network infrastructure, mobile devices, mobile device OSs, microchip design, and microchip foundries.

The United States currently has a relatively weak position in two 5G sectors—network infrastructure and microchip foundries—which are at the very top and bottom of the 5G technology stack. In the other 5G sectors, the United States has a relative advantage over or is roughly at parity with China. In the rest of this section, we provide justification for these assessments.

5G Network Infrastructure

China has a global market advantage in the 5G network infrastructure market because the United States relies on foreign suppliers to build the U.S. 5G network and because Chinese companies, Huawei and ZTE, have a significant presence in the global market. Key U.S. 5G infrastructure suppliers are European or Asian companies that might be subject to unfair competition from Chinese companies, outside of the United States and those allied countries that have not banned Huawei equipment in their 5G networks.

Mobile Devices

Chinese and U.S. companies are major players in the mobile device market. Apple is one of the largest and most-profitable companies in the world because of the success of the iPhone. In the highly competitive 5G mobile phone market, Apple and several Chinese mobile phone makers offer a variety of 5G mobile phones. Despite the Huawei blacklisting, other Chinese phone makers have a greater share of the global market than phone makers from other countries.

Mobile Device Operating Systems

Apple and Google make the two leading OSs. Most Chinese mobile device manufacturers use Google Android, which enables the United States to use Android as a lever to influence the behavior of Chinese companies. Huawei has been placed on BIS's entity list. As a consequence, Huawei's access to the secure version of the Android OS and the Google Play app store has been restricted, severely reducing its ability to produce 5G mobile phones that can securely run many popular apps outside of China. Huawei introduced its own mobile OS in 2020, but, so far, it has limited market share outside of China. Other Chinese mobile phone makers have not been blacklisted by the United States and continue to enjoy access to Android and the Google Play app store. In this 5G market segment, the United States has both market and technology advantages.

Microchip Design

U.S. chip designers are market leaders in the microchips needed for 5G networks. Qualcomm is one of the key providers of 5G modem chips. ZTE uses its chips in its base stations, and Chinese 5G phone makers use Qualcomm chipsets. In contrast, Huawei has created its own microchip design company, HiSilicon, that designs many (but not all) of the chips it needs for its 5G products. Huawei might be catching up to U.S. chip designers, but it still requires TSMC to makes its chips. The DOC blacklisting of Huawei has prevented the company from producing its chips at TSMC.

Microchip Foundries

State-of-the-art microchips are needed to meet the demanding performance requirements specified in 5G technical standards. Microchip foundries, capable of producing microchips with features that are 7 nm or smaller, are key links in the 5G supply chains of U.S. and Chinese companies. U.S. and European 5G companies must rely on Asian foundries to make 5G microchips.

There has been significant consolidation in the microchip foundry market as many players, including Intel, have encountered difficulties in manufacturing 5G chips. Only two microchip foundries—TSMC and Samsung—are presently capable of producing the high-speed logic chips needed for 5G. Until recently, TSMC supplied chips for U.S. and Chinese companies. The U.S. blacklisting of Huawei made it illegal for foreign companies to produce chips for Huawei using U.S. technology. U.S. companies provide key chip fabrication technologies to TSMC, so TSMC has complied with the U.S. blacklist order. Samsung and TSMC are currently aligned with U.S. interests because they rely on U.S. semiconductor fabrication tools in their 7-nm production lines.

However, if U.S. companies were to lose their market-leading positions in semiconductor manufacturing equipment, there is no guarantee that foreign firms, such as Samsung and TSMC, will agree to comply with U.S. blacklist orders. Even though the United States no longer has market or technology advantages in the 5G foundry sector, it has been able exert pressure on the foreign foundries supplying 5G chips to Huawei to create an advantage in this sector.

Huawei and the Impact of the U.S. Blacklisting

Table 7.2 resembles Table 7.1 but compares the positions of Huawei and the U.S. ecosystem of 5G companies. Huawei is color-coded green in the infrastructure sector because it supplies 5G networks to many foreign countries, which gives China a security advantage in those countries, enabling China to collect national security, IP, and foreign policy secrets.

The United States is green in the 5G mobile phone sector but for different reasons, because the use of U.S.-designed 5G phones reduce the Chinese government surveillance threat. Huawei is light red in this sector because the U.S. blacklisting has prevented it from selling 5G phones in most of the global market.

As before, yellow indicates that, in a specific 5G area, a particular country or company is reliant on foreign third-party suppliers and has access to the products made by these suppliers. We assumed that U.S. foreign suppliers were trustworthy and would not intentionally place cyber exploits or back doors into their products, making them less vulnerable to compromise. As before, the United States is color-coded yellow in the 5G network infrastructure and microchip foundry sectors. These are the two 5G sectors in which the United States currently has the weakest security positions.

Huawei is color-coded red or dark red in all 5G sectors, except for infrastructure. Because of the U.S. blacklisting, it has had to introduce its own mobile device OS into a mature market dominated by Apple and Google. Also because of the blacklisting, it has lost access to new and updated U.S. chip design tools and the 7-nm foundry sector.

TABLE 7.2

Huawei and U.S. 5G Market and Technology Status

5G Architecture	Huawei	United States
Network infrastructure	A_M	D_M
Mobile devices	D_M	A_T
Mobile device OS	D_M, D_T	A_M, A_T
Microchip design	D_T	A_M, A_T
Microchip foundry	D_M, D_T	D_T

NOTE: Dark green indicates that the entity has an advantage in both market and technology. Light green indicates that the entity has an advantage in either market or technology. Yellow indicates that the entity relies on foreign third-party suppliers but has access to products from these suppliers. Light red indicates that the entity has a disadvantage in either market or technology. Dark red indicates that the entity has a disadvantage in both market and technology.

Given China's loss of market share in the mobile phone market, it would not be surprising if the Chinese government has already provided financial assistance to Huawei to keep it afloat. Huawei was forced to sell its entire smartphone division, Honor, to a local Chinese government and a Chinese phone distributor, Digital China, for $15 billion, in an all-cash deal in late 2020 (Zhu, 2020). The effects of the U.S. blacklisting have provided a security advantage to the United States relative to Huawei because it has slowed the deployment of Huawei mobile devices and network infrastructure globally.

If consumers adopted Huawei's mobile OS in place of Android, the U.S. blacklisting effort might succeed only until then and could threaten the advertising business model of Google, one of the largest and most-profitable U.S. high-technology companies. It would be especially worrisome if other Chinese mobile phone manufacturers decided to drop Android for the new Huawei OS. To date, there is no evidence that such a move is in the offing, although the Chinese government could direct such a move inside China. Although it might be an effective measure in the short term, keeping the Huawei blacklisting applied to Google Android could be risky in the long term.

The United States has been able to use the chip foundry sector to weaken Huawei's position in the 5G phone market. However, the U.S. security advantage could erode quickly if China were able to build state-of-the-art microchip foundries inside China. China has probably already attempted to gain access to TSMC's technology secrets. If it could acquire this technology through cyberattack or by other means, it could establish domestic chip foundries with similar capabilities and provide Huawei a supply chain that is independent of U.S. or Taiwanese companies. However, most microchip experts consider this to be unlikely and estimate that China is a decade behind the market leaders (Kessler, 2020). In addition, without state-of-the-art semiconductor fabrication equipment from U.S. vendors, it will be even harder for China to build a domestic foundry capable of producing 5G chips.

Recommendations

A U.S. security strategy should secure the integrity of the U.S. 5G microchip supply chain and ensure the technology leadership of trusted network infrastructure providers without putting at risk the technology leadership of U.S. companies in key 5G market sectors, such as chip design, OSs, and mobile devices. We recommend that the United States undertake several activities to achieve these objectives. These are shown in Table 7.3, along with the specific government agencies recommended to carry them out.

TABLE 7.3

Current and Recommended 5G-Related Activities for Federal Entities

Organization	Function	Current 5G Activity	Recommended 5G Activity
BIS	Ensures U.S. export control and treaty compliance system in support of U.S. strategic tech leadership	Blacklist untrustworthy foreign companies. For example, BIS placed Huawei on entity list in 2019.	Continue sanctions on Huawei; permit U.S. companies to sell chips to nonsanctioned Chinese phone makers; provide incentives to establish advanced microchip foundries in the United States.
CISA	Manages cyber and physical risk to U.S. critical infrastructure	Carry out the five lines of effort specified in the CISA 5G strategy.	Monitor 5G supply chains; encourage foreign trusted vendors to adopt cybersecurity SCRM best practices.
DARPA	Basic and applied research	Discover and examine 5G network security and internet-of-things security issues.	In addition to current cybersecurity research, sponsor research on advanced semiconductor fabrication methods and tools; assist DOC by evaluating progress made by U.S. companies building 7-nm process foundries in the United States.
FCC	Regulates radio (e.g., cellular), television, wire, satellite, and cable communications	Allocate and auction Midband (C-band) spectrum for 5G networks.	License additional midband spectrum; license shared spectrum bands for 5G.
NIST	Advances science and engineering measurement standards and technology	Maintain the NIST alliance for 5G networks; discover and examine 5G cyber vulnerabilities; write and maintain 5G spectrum-sharing standards.	Continue and expand cybersecurity evaluations of 5G technical standards.
NSF	Basic research	There are no current activities.	Sponsor research on 5G and other advanced wireless communication technologies and on advanced semiconductor fabrication methods and tools.
NTIA	Advises the President on telecommunication and information policy issues	Write and maintain the 5G implementation plan for *National Strategy to Secure 5G*; conduct research on advanced civilian communication systems.	Increase budget for ITS; with assistance of DoD and appropriate FFRDCs, ITS should develop and evaluate 5G technical standards for spectrum sharing and network security.
USPTO	Grants U.S. patents, registers trademarks; advises the President, Secretary of Commerce, and federal agencies on IP policy, protection, and enforcement	Review patent applications; conduct patent quality studies.	Link USPTO patent database with the 5G 3GPP or ETSI technical standard databases; evaluate quality of Chinese 5G patents; examine links between U.S. and Chinese 5G SEP applications and application timing; provide alerts to U.S. government if 3GPP approves Chinese company–declared SEPs; propose alternative patent application disclosure rules to Congress and the WTO.

SOURCES: BIS, undated; CISA, 2020; DARPA, undated; FCC, undated; NIST, 2021; NSF, undated; NTIA, 2021; USPTO, undated.

NOTE: DARPA = Defense Advanced Research Projects Agency. FFRDC = federally funded research and development center. WTO = World Trade Organization.

Assist Trusted 5G Foreign Vendors in Securing Their Supply Chains

The United States should assist Ericsson, Nokia, and Samsung 5G to ensure the security and integrity of their infrastructure products. The U.S. ICT industry and CISA have been developing SCRM best practices for securing the supply chain of ICT products. CISA should invite these 5G companies to become members of the ICT SCRM Task Force and work with them to improve 5G SCRM.

There is ample evidence that Huawei's access to state subsidies has given it an unfair competitive advantage in global markets. Huawei and ZTE, private companies with opaque finances, could pose market risks to the two largest suppliers of U.S. 5G network infrastructure: Ericsson and Nokia. The financial health of these two critical U.S. 5G foreign suppliers should be monitored by the U.S. government.

Allocate More Midband Spectrum to 5G and Investigate Spectrum Sharing

The FCC should allocate more spectrum at midband frequencies to 5G networks. Some midband spectrum is used by government agencies. Spectrum sharing might be feasible in these bands and could reduce the cost of spectrum for U.S. carriers. DoD and ITS have developed or examined spectrum-sharing technologies and standards. 3GPP is also working on spectrum-sharing standards. A concerted and coordinated effort to develop 5G spectrum-sharing standards that will work in the U.S. midband spectrum would be beneficial to many U.S. industries.

Constrain Huawei's Access to Advanced 5G Chips

Huawei can no longer build 5G phones and reportedly has sold its mobile phone division (although, given its opaque finances, this might, in reality, be just a government bailout). Huawei is reportedly dependent on Intel and Xilinx chips for its 5G base stations (Bosnjak, 2020). In recognition of the threat to its business, Huawei announced that it had purchased more than a two-year supply of U.S. chips to ensure that it can produce 5G base stations (Li and Ting-Fang, 2020). Huawei will need to procure additional 5G chips when its stockpile is exhausted. There are reports that Huawei has attempted to secure an alternative source of 5G chips from Samsung in South Korea, but it might not have been able to do so (Chedalla, 2020). By keeping Huawei on the DOC entity list, the United States will slowly hamper Chinese intelligence collection and IP theft operations around the world.

Exclude U.S. 5G Chip Suppliers from Sanctions

BIS should carefully reconsider the companies to which U.S. chip designers can sell their chips. Qualcomm, a U.S. company, provides an important technology leadership position for the United States in the 5G architecture. To a lesser extent, so do Intel and Apple. The U.S. government has forbidden Qualcomm to sell chips to Huawei. However, Qualcomm does sell chips to other Chinese 5G companies and, in particular, to mobile phone manufacturers. It is in the U.S. interest to allow Qualcomm to still sell chips to Chinese phone makers that do not provide infrastructure and that have not demonstrated a history of espionage. We recommend that Qualcomm still be permitted to sell chips to Chinese phone makers, but the United States must ensure that those chips do not leak into Huawei's supply chain. It might be possible to track Qualcomm chips in the supply chain of Chinese companies if trusted platform module circuitry is embedded in these chips. These chips could be programmed to report their device characteristics when they are first activated. This might make it possible to detect U.S. chips that are activated for service in Huawei devices.

Permitting Qualcomm, Skyworks Solutions, Qorvo, and other U.S. chipmakers to sell chips to Chinese phone makers helps ensure the economic health of the U.S. 5G supply chain and reduces the economic incentives for the Chinese government to support the development of advanced chipmaking capabilities in China.

It would also help dissociate current U.S. policy toward Huawei from larger, unresolved trade issues between the United States and China.

Protect Technology in U.S. Patent Applications and Examine Chinese 5G Patents

U.S. policymakers require a better understanding of where and how U.S. and Chinese company technologies are incorporated into 5G standards. To do this, we recommend that USPTO link its patent database with the 5G 3GPP or ETSI technical standard databases. This will enable patent experts to evaluate the quality of Chinese 5G patents submitted as 5G SEPs, identify competing U.S. patent claims, examine the timing of SEP applications, and identify specific areas of technology competition between China and the United States and technology areas in which China might be using unethical or illegal methods to acquire U.S. technology.

The USPTO should also monitor the growing competition between U.S. and Chinese firms in SEP submissions and 3GPP adjudications of submitted SEP declarations. The USPTO should identify overlapping 5G SEPs declared by Chinese and Western firms, the dates such patents are filed, where they are filed, SEP quality, and SEP approval, to provide U.S. policymakers indications of whether there is a shift in intellectual leadership in communication technologies from Western to Chinese companies. This information can guide the development of new U.S. government–sponsored research into such technologies, to ensure that U.S. companies retain technical leadership in critical areas of wireless communications.

Some experts contend that Chinese companies review U.S. patent applications as soon as they are put online and use this information to file their own competing patents in China, and possibly in other countries. This enables China to reduce competition in its domestic market for 5G systems, could slow the approval of declared 5G SEPs filed by U.S. companies, and threatens the integrity of the global patent system. This practice might also have enabled Chinese companies to file the large number of 5G that they have in the past few years. The USPTO should ask Congress to keep patent applications secret until they are approved or until just shortly before approval, as was the case for many decades prior to 1994. This change would also require a change to WTO regulations, so the administration should immediately raise this issue with the WTO as well.

Establish a U.S. Microchip Fabrication Research and Development Program

Both U.S. and Chinese companies remain dependent on only a few key suppliers of the most-advanced 5G chips—namely, TSMC and Samsung. Importantly, the United States and Advanced Semiconductor Materials Lithography in the Netherlands provide the most-advanced tools used in chip manufacturing (Kharpal, 2019). NSF should sponsor research on advanced microchip manufacturing technologies. The United States, with the assistance of its allies, needs to retain leadership in these areas in order to control access to the supply chain for advanced microchips. NSF should also fund research in EUV microchip lithography. This critical technology will enable U.S. chipmakers catch up to the Asian industry leaders. This will be expensive, and it might be possible to build only one or two EUV R&D facilities in the United States, so NSF should ensure that a wide variety of U.S. researchers, government agencies, and U.S. companies are granted access to these facilities.

Establish State-of-the-Art Microchip Foundries in the United States

U.S. efforts to encourage TSMC and Samsung to build foundries in the United States should be continued (Davis, O'Keeffe, and Fitch, 2020). At least one foreign-owned foundry should be built in the United States because this will ensure a trusted source of advanced microchips in case U.S. chipmakers falter.

Although the United States has many first-in-class microchip design firms, it has only two remaining microchip manufacturing companies that are near the state of the art: Intel and GlobalFoundries. For many years, Intel led the world in the design and production of computer microprocessors. However, its leadership role could be eclipsed by AMD, a fabless chip designer that has its chips made by TSMC. Intel has recently taken steps to adopt leading-edge EUV lithography to catch up to microchip leaders. GlobalFoundries has recently become again a U.S. firm but one that has fallen behind the market leaders. In contrast to Intel, it has not indicated whether it will try to once again become a market leader. Either Intel or both Intel and Global-Foundries will need additional help to produce leading-edge microchips. It would be prudent for local, state, and federal government bodies to provide them financial incentives to build new state-of-the-art foundries in the United States if they ask for assistance. These incentives would include those outlined in the proposed CHIPS Act and carried out by the Secretary of Commerce.

Develop Requirements for Huawei to Enable Its Return to 5G Markets

We recommend that DOC, in consultation with the U.S. National Security Council, develop a set of conditions for Huawei to meet to demonstrate that it can become a trustworthy partner for 5G. Huawei meeting these conditions would enable U.S. chip designers and foreign foundries to sell their products and services to these sanctioned companies. It would also defuse tensions between the United States and China. These conditions should explicitly prohibit IP theft, espionage, and the violation of U.S. sanctions against North Korea or Iran.

These recommendations are consistent the high-level goals of the current U.S. *National Strategy to Secure 5G* and the associated NTIA implementation plan. To achieve these goals, a whole-of-government approach is needed that builds on the existing authorities and capabilities of U.S. agencies.

Areas of Potential Future Research

In this report, we have identified several areas in which further research is needed to help secure U.S. 5G infrastructure and ensure that a U.S. 5G security strategy is implemented effectively.

Chinese companies have filed a large number of patents and have submitted these to 3GPP as declared SEPs. Companies from both the United States and China have expended significant resources into the development of cellular communications and related advanced technologies. The U.S. government should track these technology developments and determine which have significant implications for advanced communications. This is highly technical work that several FFRDCs could pursue.

An important question is whether a government agency or new private organization should be put in charge of a new microchip fabrication R&D program, how much government funding should be allocated to this effort, and whether and how costs should be shared with industry. At one time, the U.S. government established a private organization called Semiconductor Manufacturing Technology (SEMATECH) to do largely the same thing, initially partly funded by DoD. However, government funding was eventually eliminated, and SEMATECH evolved into a private, internationally funded R&D organization, which might not be the model now needed to achieve U.S. government objectives. A follow-on study could examine what sort of organization should lead this future R&D effort.

Exactly what conditions Huawei should be required to meet to gain access to the 5G microchip supply chain should be the subject of future research and could require significant changes to Huawei equipment. At a minimum, these requirements should include Huawei rescinding its internal policies that reward IP theft by employees, full disclosure of all subsidies received from the Chinese government, the return of state subsidies that allow it to compete unfairly in 5G markets, and changes in its corporate governance structure.

Abbreviations

3G	third generation
3GPP	3rd Generation Partnership Project
4G	fourth generation
5G	fifth generation
6G	sixth generation
AMD	Advanced Micro Devices
BIS	Bureau of Industry and Security
CCP	Chinese Communist Party
CHIPS Act	Creating Helpful Incentives to Produce Semiconductors for America Act
CISA	Cybersecurity and Infrastructure Security Agency
CNIPA	China National Intellectual Property Administration
COVID-19	coronavirus disease 2019
DOC	U.S. Department of Commerce
DoD	U.S. Department of Defense
EIB	European Investment Bank
EPO	European Patent Office
ETSI	European Telecommunications Standards Institute
E-UTRA	evolved universal terrestrial radio access
EUV	extreme ultraviolet
FCC	Federal Communications Commission
FFRDC	federally funded research and development center
GHz	gigahertz
ICT	information and communication technology
IDM	integrated device manufacturer
IoT	internet of things
IP	intellectual property
IPR	intellectual property rights
ITS	Institute for Telecommunication Sciences
LTE	long-term evolution
MAC	medium access control
NG-RAN	next-generation radio access network
NIST	National Institute of Standards and Technology
nm	nanometer
NR	new radio
NSF	National Science Foundation
NTIA	National Telecommunications and Information Administration
OS	operating system
PDCP	packet data convergence protocol

RAN	radio access network
R&D	research and development
RFFE	radio frequency front end
RLC	radio link control
RRC	radio resource control
SCRM	supply chain risk management
SDN	software-defined network
SEP	standard-essential patent
SMIC	Semiconductor Manufacturing International Corporation
SSO	standard-setting organization
TSG	technical specification group
TSMC	Taiwan Semiconductor Manufacturing Company
UDF	unified data function
UDM	unified data management
UE	user equipment
UK	United Kingdom
UMC	United Microelectronics Corporation
URLLC	ultrareliable and low-latency communication
USPTO	U.S. Patent and Trademark Office
V2X	vehicle to everything
WG	working group
WIPO	World Intellectual Property Organization
WTO	World Trade Organization

References

3GPP, homepage, undated a. As of June 11, 2020:
https://www.3gpp.org/

———, "3GPP Officials per TSG/WG: 3GPP Officials for Group: 3GPP RAN 1 ('R1')," undated b. As of June 2, 2020:
https://www.3gpp.org/DynaReport/TSG-WG--R1--officials.htm

———, "3GPP Officials per TSG/WG: 3GPP Officials for Group: 3GPP RAN 2 ('R2')," undated c. As of June 2, 2020:
https://www.3gpp.org/DynaReport/TSG-WG--R2--officials.htm

———, "3GPP Officials per TSG/WG: 3GPP Officials for Group: 3GPP RAN 3 ('R3')," undated d. As of June 2, 2020:
https://www.3gpp.org/DynaReport/TSG-WG--R3--officials.htm

———, "3GPP Officials per TSG/WG: 3GPP Officials for Group: 3GPP RAN 4 ('R4')," undated e. As of June 2, 2020:
https://www.3gpp.org/DynaReport/TSG-WG--R4--officials.htm

———, "3GPP Officials per TSG/WG: 3GPP Officials for Group: 3GPP RAN 5 ('R5')," undated f. As of June 2, 2020:
https://www.3gpp.org/DynaReport/TSG-WG--R5--officials.htm

———, "3GPP Officials for Group: R6," undated g. As of June 2, 2020:
https://www.3gpp.org/DynaReport/TSG-WG--R6--officials.htm

———, "3GPP Officials for Group: 3GPP RAN ('RP')," undated h. As of June 2, 2020:
https://www.3gpp.org/DynaReport/TSG-WG--RP--officials.htm

———, "About 3GPP Home," undated i. As of December 5, 2021:
https://www.3gpp.org/about-3gpp/about-3gpp

———, "Legal Matters," undated j. As of September 18, 2021:
https://www.3gpp.org/about-3gpp/legal-matters

———, "Release 16," undated k. As of June 11, 2020:
https://www.3gpp.org/release-16

———, "Releases," undated l. As of August 29, 2021:
https://www.3gpp.org/specifications/releases

———, "Specifications Groups Home," undated m. As of June 2, 2020:
https://www.3gpp.org/specifications-groups/specifications-groups

———, "Release 17," December 12, 2020. As of August 29, 2021:
https://www.3gpp.org/release-17

———, "3GPP Working Procedures," webpage, effective April 29, 2021a. As of April 30, 2020:
https://www.3gpp.org/ftp/Information/Working_Procedures/3GPP_WP.htm

———, "RAN1—Radio Layer 1 (Physical Layer)," webpage, updated October 15, 2021b. As of January 19, 2022:
https://www.3gpp.org/specifications-groups/ran-plenary/ran1-radio-layer-1

5G Infrastructure Public Private Partnership Architecture WG—*See* 5G Infrastructure Public Private Partnership Architecture Working Group.

5G Infrastructure Public Private Partnership Architecture Working Group, *View on 5G Architecture*, version 3.0, Belgium: European Commission, June 19, 2019. As of June 11, 2020:
https://5g-ppp.eu/wp-content/uploads/2019/07/
5G-PPP-5G-Architecture-White-Paper_v3.0_PublicConsultation.pdf

Adee, Sally, "The Hunt for the Kill Switch," *IEEE Spectrum*, Vol. 45, No. 5, May 2008, pp. 34–39.

Alcorn, Paul, "Intel's 7nm Is Broken, Company Announces Delay Until 2022, 2023," *Tom's Hardware*, July 23, 2020. As of July 24, 2020:
https://www.tomshardware.com/news/intel-announces-delay-to-7nm-processors-now-one-year-behind-expectations

Alsop, Thomas, "Qualcomm R&D; Expenditures Worldwide 2016–2020," Statista, April 29, 2021.

Apple, "Apple to Acquire the Majority of Intel's Smartphone Modem Business," press release, Cupertino and Santa Clara, Calif., July 25, 2019. As of June 1, 2020:
https://www.apple.com/newsroom/2019/07/apple-to-acquire-the-majority-of-intels-smartphone-modem-business/

Armasu, Lucian, "Report: TSMC to Start 3nm Volume Production in 2022," *Tom's Hardware*, December 6, 2019. As of May 21, 2020:
https://www.tomshardware.com/news/report-tsmc-to-start-3nm-volume-production-in-2022

Bajpai, Prableen, "Which Companies Spend the Most in Research and Development (R&D)?" Nasdaq, June 21, 2021. As of December 6, 2021:
https://www.nasdaq.com/articles/which-companies-spend-the-most-in-research-and-development-rd-2021-06-21

Barfield, Claude, "Nationalization: The Answer on 5G—or Just Evidence of U.S. Flailing in the Face of the China Challenge?" blog post, *AEIdeas*, March 20, 2019. As of June 11, 2020:
https://www.aei.org/technology-and-innovation/telecommunications/nationalization-the-answer-on-5g-or-just-evidence-of-us-flailing-in-the-face-of-the-china-challenge/

Baron, Justus, and Kirti Gupta, "Unpacking 3GPP Standards," *Journal of Economics and Management Strategy*, Vol. 27, No. 3, September 2018, pp. 433–461.

BIS—*See* Bureau of Industry and Security.

Bonds, Timothy M., James Bonomo, Daniel Gonzales, C. Richard Neu, Samuel Absher, Edward Parker, Spencer Pfeifer, Jennifer Brookes, Julia Brackup, Jordan Willcox, David R. Frelinger, and Anita Szafran, *America's 5G Era: Gaining Competitive Advantages While Securing the Country and Its People*, Santa Monica, Calif.: RAND Corporation, PE-A435-1, 2021. As of December 5, 2021:
https://www.rand.org/pubs/perspectives/PEA435-1.html

Bosnjak, Dominik, "Concerns Mount as Huawei's Chip Stockpile Grows," blog post, *Android Headlines*, May 28, 2020. As of June 11, 2020:
https://www.androidheadlines.com/2020/05/concerns-mount-as-huaweis-chip-stockpile-grows.html

Brackup, Julia, Sarah Harting, and Daniel Gonzales, *Digital Infrastructure and Digital Presence: A Framework for Assessing the Impact on Future Military Competition and Conflict*, Santa Monica, Calif.: RAND Corporation, in production.

Brewster, Thomas, "U.S. Funds Program with Free Android Phones for the Poor—but with Permanent Chinese Malware," *Forbes*, January 9, 2020. As of December 7, 2021:
https://www.forbes.com/sites/thomasbrewster/2020/01/09/us-funds-free-android-phones-for-the-poor---but-with-permanent-chinese-malware/

Bureau of Industry and Security, U.S. Department of Commerce, "About BIS," undated. As of December 5, 2021:
https://www.bis.doc.gov/index.php/about-bis

Byford, Sam, "Huawei's Mate 30 Contains No American Parts," *The Verge*, December 3, 2019. As of June 11, 2020:
https://www.theverge.com/2019/12/3/20993148/huawei-mate-30-no-american-components-manufacturers

CableLabs, "A Comparative Introduction to 4G and 5G Authentication," blog post, *Inform[ED] Insights*, Winter 2019. As of June 4, 2020:
https://www.cablelabs.com/insights/a-comparative-introduction-to-4g-and-5g-authentication

Cain, Geoffrey, *Samsung Rising: The Inside Story of the South Korean Giant That Set Out to Beat Apple and Conquer Tech*, New York: Currency, 2020.

Chedalla, Teja, "Huawei Looking at a Potential Chipset Deal with Korean Companies," blog post, *Android Headlines*, June 2, 2020. As of June 11, 2020:
https://www.androidheadlines.com/2020/06/
huawei-looking-at-a-potential-chipset-deal-with-korean-companies.html

Chin, Spencer, "Top-Tier Smartphone Makers Going to In-House Processors: Report," *FierceElectronics*, January 9, 2020. As May 20, 2020:
https://www.fierceelectronics.com/electronics/top-tier-smartphone-makers-going-to-house-processors-report

CHIPS Act—*See* U.S. House of Representatives, 2020, and U.S. Senate, 2020.

Choate, Pat, Manufacturing Policy Project, *A Great Wall of Patents: China and American Investors—Selected Consequences of Proposed U.S. Patent "Reforms,"* working paper prepared for the U.S.–China Economic and Security Review Commission, November 7, 2005. As of December 6, 2021:
https://www.uscc.gov/research/
working-paper-great-wall-patents-china-and-american-investors-selected-consequences

CISA—*See* Cybersecurity and Infrastructure Security Agency.

Cisco, *Wi-Fi 6 and Private LTE/5G Technology and Business Models in Industrial IoT: Wireless Trends and the Different Phases of Technology Adoption*, white paper, 2019.

Clancy, T. Charles, executive director, Hume Center for National Security and Technology and Bradley Professor of Cybersecurity, Virginia Polytechnic Institute and State University, testimony before the U.S. Senate Committee on the Judiciary, hearing titled "5G: National Security Concerns, Intellectual Property Issues, and the Impact on Competition and Innovation," Washington, D.C., May 14, 2019. As of December 5, 2021:
https://www.judiciary.senate.gov/meetings/
5g-national-security-concerns-intellectual-property-issues-and-the-impact-on-competition-and-innovation

Clark, Robert, "No One Wants to Talk About Huawei's State Subsidies," *Light Reading*, January 9, 2020. As of June 11, 2020:
https://www.lightreading.com/asia-pacific/no-one-wants-to-talk-about-huaweis-state-subsidies/d/d-id/756697

Contreras, Jorge L., "Essentiality and Standards-Essential Patents," in Jorge L. Contreras, ed., *The Cambridge Handbook of Technical Standardization Law: Competition, Antitrust, and Patents*, Cambridge: Cambridge University Press, 2017, pp. 209–230.

Conway, Adam, "The 7nm Qualcomm Snapdragon X55 5G Modem Has Been Announced Ahead of MWC," XDA-Developers, February 19, 2019. As of May 21, 2020:
https://www.xda-developers.com/qualcomm-snapdragon-x55-5g-modem-2019-android-smartphones/

Cornell University, Institut Européen d'Administration des Affaires, and WIPO—*See* Cornell University, Institut Européen d'Administration des Affaires, and World Intellectual Property Organization.

Cornell University, Institut Européen d'Administration des Affaires, and World Intellectual Property Organization, *The Global Innovation Index 2020: Who Will Finance Innovation?* Ithaca, Fontainebleau, and Geneva, 2020. As of December 6, 2021:
https://www.wipo.int/global_innovation_index/en/2020

Counterpoint, "Global Smartphone Market Share: By Quarter," November 29, 2021. As of December 6, 2021:
https://www.counterpointresearch.com/global-smartphone-share/

Crawford, Susan, "America Needs More Fiber," *Wired*, February 8, 2018. As of May 21, 2020:
https://www.wired.com/story/america-needs-more-fiber/

Cybersecurity and Infrastructure Security Agency, U.S. Department of Homeland Security, "Information and Communications Technology (ICT) Supply Chain Risk Management (SCRM) Task Force," undated. As of December 7, 2021:
https://www.cisa.gov/ict-scrm-task-force

———, *CISA 5G Strategy: Ensuring the Security and Resilience of 5G Infrastructure in Our Nation*, 2020. As of December 5, 2021:
https://www.cisa.gov/publication/5g-strategy

Dahir, Abdi Latif, "China 'Gifted' the African Union a Headquarters Building and Then Allegedly Bugged It for State Secrets," *Quartz Africa*, January 30, 2018. As of June 11, 2020:
https://qz.com/africa/1192493/china-spied-on-african-union-headquarters-for-five-years/

Dano, Mike, "Attorney General Barr: The US, Allies Should Take Nokia or Ericsson Stake for 5G," *Light Reading*, February 6, 2020. As of June 4, 2020:
https://www.lightreading.com/mobile/5g/
attorney-general-barr-the-us-allies-should-take-nokia-or-ericsson-stake-for-5g/d/d-id/757341

DARPA—*See* Defense Advanced Research Projects Agency.

DataReportal, "Digital Around the World," undated. As of May 16, 2020:
https://datareportal.com/global-digital-overview

Davis, Bob, Kate O'Keeffe, and Asa Fitch, "Taiwan Firm to Build Chip Factory in U.S.," *Wall Street Journal*, May 14, 2020.

Davis, Bob, Dan Strumpf, and Lingling Wei, "China's ZTE to Pay $1 Billion Fine in Settlement with U.S.," *Wall Street Journal*, June 7, 2018.

Defense Advanced Research Projects Agency, "About DARPA," webpage, undated. As of March 10, 2022:
https://www.darpa.mil/about-us/about-darpa

Diwakar, Amar, "'Chip Wars': US, China and the Battle for Semiconductor Supremacy," *TRT World*, March 16, 2021. As of December 7, 2021:
https://www.trtworld.com/magazine/chip-wars-us-china-and-the-battle-for-semiconductor-supremacy-45052

Downes, Larry, "The U.S. Government Shouldn't Run the Country's 5G Network," *Harvard Business Review*, April 30, 2019.

Ericsson, *5G Security: Enabling a Trustworthy 5G System*, March 28, 2018, republished March 2021. As of June 11, 2020:
https://www.ericsson.com/en/reports-and-papers/white-papers/5g-security---enabling-a-trustworthy-5g-system

ETSI—*See* European Technical Standards Institute.

European Technical Standards Institute, "ETSI IPR Online Database," webpage, undated. As of January 19, 2022:
https://ipr.etsi.org/

FCC—*See* Federal Communications Commission.

Federal Communications Commission, "What We Do," undated. As of December 5, 2021:
https://www.fcc.gov/about-fcc/what-we-do

Freifeld, Karen, and Eric Auchard, "U.S. Probing Huawei for Possible Iran Sanctions Violations: Sources," Reuters, April 25, 2018. As of December 5, 2021:
https://www.reuters.com/article/us-usa-huawei-doj/
u-s-probing-huawei-for-possible-iran-sanctions-violations-sources-idUSKBN1HW1YG

Gargeyas, Arjun, "China's 'Standards 2035' Project Could Result in a Technological Cold War," *The Diplomat*, September 18, 2021. As of December 6, 2021:
https://thediplomat.com/2021/09/chinas-standards-2035-project-could-result-in-a-technological-cold-war/

Gartenberg, Chaim, "Intel Says Apple and Qualcomm's Surprise Settlement Pushed It to Exit Mobile 5G," *The Verge*, April 25, 2019a. As of June 11, 2020:
https://www.theverge.com/2019/4/25/18516830/
intel-apple-qualcomm-surprise-settlement-pushed-exit-mobile-5g-modems

———, "Qualcomm Will Get At Least $4.5 Billion from Apple as Part of Its Patent Settlement," *The Verge*, May 1, 2019b. As of June 11, 2020:
https://www.theverge.com/2019/5/1/18525962/qualcomm-apple-settlement-4-5-billion-dollars-patent

Gillen, Brad, "In the Global 5G Race, America's Free Market Leads the Way," blog post, CTIA, February 22, 2019. As of May 21, 2020:
https://www.ctia.org/news/blog-in-the-global-5g-race-americas-free-market-leads-the-way

Griffith, Keith, "Nokia to Acquire Alcatel-Lucent for $16.6 Billion," *SDxCentral*, April 15, 2015. As of June 11, 2020:
https://www.sdxcentral.com/articles/news/nokia-to-acquire-alcatel-lucent-for-16-6b/2015/04/

"Huawei Mate 30 Pro Teardown," *iFixit*, November 14, 2019. As of June 11, 2020:
https://www.ifixit.com/Teardown/Huawei+Mate+30+Pro+Teardown/127743

Huawei, "Who Is Huawei," webpage, undated. As of July 30, 2020:
https://www.huawei.com/us/corporate-information

Indrayan, Gunjan, "Wireless Telecom Infrastructure Market Worldwide: Trends and Developments," forum post, *Wireless and Mobile Telecommunications*, September 2, 2014. As of June 11, 2020:
https://wirelesstelecom.wordpress.com/2014/09/02/
wireless-telecom-infrastructure-market-worldwide-trends-and-developments/

Intel, "Intel Reports Fourth-Quarter and Full-Year 2020 Financial Results," press release, January 21, 2021. As of December 6, 2021:
https://www.intc.com/news-events/press-releases/detail/1439/
intel-reports-fourth-quarter-and-full-year-2020-financial

Ip, Greg, "China's Rise Drives a U.S. Experiment in Industrial Policy," *Wall Street Journal*, March 10, 2021.

Jowitt, Tom, "Dutch Report Flagged Huawei Monitoring of KPN Concerns—Report," *Silicon*, April 21, 2021a. As of December 7, 2021:
https://www.silicon.co.uk/5g/dutch-report-huawei-kpn-monitoring-393727

———, "Lithuania Warns Citizens over 5G Chinese Phones," *Silicon*, September 22, 2021b. As of December 7, 2021:
https://www.silicon.co.uk/mobility/smartphones/lithuania-warns-citizens-over-5g-chinese-phones-417535

Kapko, Matt, "Nokia Stares Down Hostile Takeover Bid," *SDxCentral*, April 17, 2020. As of June 11, 2020:
https://www.sdxcentral.com/articles/news/nokia-stares-down-hostile-takeover-bid/2020/04/

Kelion, Leo, "Huawei Set for Limited Role in UK 5G Networks," BBC News, January 28, 2020a. As of June 11, 2020:
https://www.bbc.com/news/technology-51283059

———, "Huawei 5G Kit Must Be Removed from UK by 2027," BBC News, July 14, 2020b. As of December 5, 2021:
https://www.bbc.com/news/technology-53403793

Kessler, Andy, "China Is Losing Its Bet on Chips," *Wall Street Journal*, November 15, 2020.

Kharpal, Arjun, "China's Biggest Chipmaker Is Still Years Behind Its Global Rivals," CNBC, August 5, 2019. As of June 11, 2020:
https://www.cnbc.com/2019/08/06/smic-chinas-biggest-chipmaker-is-still-years-behind-its-rivals.html

———, "The Extradition Trial of Huawei's CFO Starts This Month—Here's What to Watch," CNBC, January 9, 2020a. As of June 11, 2020:
https://www.cnbc.com/2020/01/10/huawei-cfo-meng-wanzhou-extradition-trial-explained.html

———, "Why New U.S. Rules on Selling Chips to Huawei Could Be a 'Big Blow' for the Chinese Tech Giant," CNBC, May 18, 2020b. As of June 11, 2020:
https://www.cnbc.com/2020/05/18/huawei-faces-big-blow-from-new-us-rules-to-cut-off-chips.html

Lepido, Daniele, "Vodafone Found Hidden Backdoors in Huawei Equipment," Bloomberg, April 30, 2019. As of December 15, 2019:
https://www.bloomberg.com/news/articles/2019-04-30/vodafone-found-hidden-backdoors-in-huawei-equipment

Lewis, James Andrew, *How Will 5G Shape Innovation and Security: A Primer*, Center for Strategic and International Studies, December 6, 2018. As of June 11, 2020:
https://www.csis.org/analysis/how-5g-will-shape-innovation-and-security

———, senior vice president, Technology Policy Program, Center for Strategic and International Studies, testimony before the U.S. Senate Committee on the Judiciary hearing titled "5G: National Security Concerns, Intellectual Property Issues, and the Impact on Competition and Innovation," Washington, D.C., May 14, 2019. As of January 19, 2022:
https://www.judiciary.senate.gov/meetings/
5g-national-security-concerns-intellectual-property-issues-and-the-impact-on-competition-and-innovation

Li, Lauly, and Cheng Ting-Fang, "Huawei Builds Up 2-Year Reserve of 'Most Important' US Chips," *Nikkei Asian Review*, May 28, 2020.

Martin, Timothy W., and Eva Dou, "Global Telecom Carriers Attacked by Suspected Chinese Hackers," *Wall Street Journal*, June 24, 2019.

Mayersen, Isaiah, "AMD Is Set to Become TSMC's Biggest 7nm Customer in 2020," *TechSpot*, January 4, 2020. As of May 21, 2020:
https://www.techspot.com/news/83400-amd-set-become-tsmc-biggest-7nm-customer-2020.html

McGill, Margaret Harding, "Republicans, Industry Shun Idea of Nationalized 5G Network," *Politico*, January 29, 2018. As of May 21, 2020:
https://www.politico.com/story/2018/01/29/nationalized-5g-network-republicans-industry-reaction-314978

McGregor, Jim, "Globalfoundries' Change in Strategy Pays Off," *Forbes*, September 17, 2019.

Medin, Milo, and Gilman Louie, *The 5G Ecosystem: Risks and Opportunities for DoD*, Defense Innovation Board, April 3, 2019. As of December 5, 2021:
https://media.defense.gov/2019/Apr/03/2002109302/-1/-1/0/DIB_5G_STUDY_04.03.19.PDF

Morgenson, Gretchen, and Tom Winter, "The U.S. Is Now Investigating Chinese Telecom Giant ZTE for Alleged Bribery," NBC News, March 13, 2020. As of December 6, 2021:
https://www.nbcnews.com/business/corporations/
u-s-now-investigating-chinese-telecom-giant-zte-alleged-bribery-n1156696

Morris, Anne, "Nokia Bolsters 5G R&D Coffers with $283 Million Loan," *SDxCentral*, December 3, 2018. As of June 11, 2020:
https://www.sdxcentral.com/articles/news/nokia-bolsters-5g-rd-coffers-with-283-million-loan/2018/12/

Morris, Iain, "Nokia Hires Marvell to Fix 5G Problems," *Light Reading*, March 4, 2020. As of July 26, 2020:
https://www.lightreading.com/5g/nokia-hires-marvell-to-fix-5g-problems/d/d-id/757984

National Academies of Sciences, Engineering, and Medicine, *Telecommunications Research and Engineering at the Institute for Telecommunication Sciences of the Department of Commerce: Meeting the Nation's Telecommunications Needs*, Washington, D.C.: National Academies Press, 2015. As of December 6, 2021:
https://doi.org/10.17226/21867

———, *The Transformational Impact of 5G: Proceedings of a Workshop–in Brief*, Washington, D.C.: National Academies Press, 2019. As of December 6, 2021:
https://doi.org/10.17226/25598

National Cyber Security Centre, Ministry of National Defence, Republic of Lithuania, *Assessment of Cybersecurity of Mobile Devices Supporting 5G Technology Sold in Lithuania: Analysis of Products Made by Huawei, Xiaomi and OnePlus*, Vilnius, August 23, 2021. As of December 7, 2021:
https://www.nksc.lt/doc/en/analysis/2021-08-23_5G-CN-analysis_env3.pdf

National Institute of Standards and Technology, U.S. Department of Commerce, "NIST Mission, Vision, Core Competencies, and Core Values," created July 10, 2009, updated March 4, 2021. As of December 5, 2021:
https://www.nist.gov/about-nist/our-organization/mission-vision-values

National Research Council, *Renewing U.S. Telecommunications Research*, Washington, D.C.: National Academies Press, 2006. As of December 6, 2021:
https://doi.org/10.17226/11711

National Science Foundation, "About the National Science Foundation," webpage, undated. As of March 10, 2022:
https://www.nsf.gov/about/

———, "Protecting Our Processors," news release 14-125, September 23, 2014. As of December 9, 2021:
https://www.nsf.gov/news/news_summ.jsp?cntn_id=132795

National Telecommunications and Information Administration, U.S. Department of Commerce, "National Strategy to Secure 5G Implementation Plan," January 19, 2021. As of December 5, 2021:
https://www.ntia.gov/5g-implementation-plan

Nellis, Stephen, "Intel to Spend $20 Billion on U.S. Chip Plants as CEO Challenges Asia Dominance," Reuters, March 23, 2021. As of December 6, 2021:
https://www.reuters.com/world/asia-pacific/
intel-doubles-down-chip-manufacturing-plans-20-billion-new-arizona-sites-2021-03-23/

Nichols, Shaun, "What the Cell . . . ? Telcos Around the World Were So Severely Pwned, They Didn't Notice the Hackers Setting Up VPN Points," *The Register*, June 25, 2019. As of June 11, 2020:
https://www.theregister.co.uk/2019/06/25/global_telcos_hacked/

NIST—*See* National Institute of Standards and Technology.

NSF—*See* National Science Foundation.

NTIA—*See* National Telecommunications and Information Administration.

Nunno, Richard, International Bureau, Federal Communications Commission, "Migration to 3G Technology Standards: A Comparison of Selected Countries," September 2003. As of December 6, 2021:
https://docs.fcc.gov/public/attachments/DOC-239506A1.pdf

Nuttall, Chris, "UK Gives 5G OK to Huawei," *Financial Times*, January 28, 2020.

Office of Public Affairs, U.S. Department of Justice, "ZTE Corporation Agrees to Plead Guilty and Pay Over $430.4 Million for Violating U.S. Sanctions by Sending U.S.-Origin Items to Iran," news release, March 7, 2017. As of December 6, 2021:
https://www.justice.gov/opa/pr/
zte-corporation-agrees-plead-guilty-and-pay-over-4304-million-violating-us-sanctions-sending

———, "PRC State-Owned Company, Taiwan Company, and Three Individuals Charged with Economic Espionage," news release, November 1, 2018. As of June 11, 2020:
https://www.justice.gov/opa/pr/
prc-state-owned-company-taiwan-company-and-three-individuals-charged-economic-espionage

———, "Chinese Telecommunications Device Manufacturer and Its U.S. Affiliate Indicted for Theft of Trade Secrets, Wire Fraud, and Obstruction of Justice," news release, January 28, 2019. As of December 15, 2019:
https://www.justice.gov/opa/pr/
chinese-telecommunications-device-manufacturer-and-its-us-affiliate-indicted-theft-trade

Omdia, "Global 4G LTE Infrastructure Revenue Totaled $22.9 Billion in 2018," analyst opinion, April 3, 2019. As of May 16, 2020:
https://technology.informa.com/612559/global-4g-lte-infrastructure-revenue-totaled-229-billion-in-2018

Palagummi, Phanidra, Vedant Somani, Krishna M. Sivalingam, and Balaji Venkat, "An Overview of the 5G Mobile Network Architecture," *Advanced Computing and Communications*, Vol. 2, No. 3, September 2018. As of June 4, 2020:
https://acc.digital/an-overview-of-the-5g-mobile-network-architecture/3/

Pancevski, Bojan, "U.S. Officials Say Huawei Can Covertly Access Telecom Networks," *Wall Street Journal*, updated February 12, 2020.

Patent Cooperation Treaty, Albania, Algeria, Angola, Antigua and Barbuda, Armenia, Australia, Austria, Azerbaijan, Bahrain, Barbados, Belarus, Belgium, Belize, Benin, Bosnia and Herzegovina, Botswana, Brazil, Brunei Darussalam, Bulgaria, Burkina Faso, Cambodia, Cameroon, Canada, Central African Republic, Chad, Chile, China, Colombia, Comoros, Congo, Costa Rica, Côte d'Ivoire, Croatia, Cuba, Cyprus, Czechia, Democratic People's Republic of Korea, Denmark, Djibouti, Dominica, Dominican Republic, Ecuador, Egypt, El Salvador, Equatorial Guinea, Estonia, Eswatini, Finland, France, Gabon, Gambia, Ghana, Georgia, Germany, Greece, Grenada, Guatemala, Guinea, Guinea-Bissau, Honduras, Hungary, Iceland, India, Indonesia, Ireland, Islamic Republic of Iran, Israel, Italy, Jamaica, Japan, Jordan, Kazakhstan, Kenya, Kuwait, Kyrgyzstan, Lao People's Democratic Republic, Latvia, Lesotho, Liberia, Libya, Liechtenstein, Lithuania, Luxembourg, Madagascar, Malawi, Malaysia, Mali, Malta, Mauritania, Mexico, Monaco, Mongolia, Montenegro, Morocco, Mozambique, Namibia, Netherlands, New Zealand, Nicaragua, Niger, Nigeria, North Macedonia, Norway, Oman, Panama, Papua New Guinea, Peru, Philippines, Poland, Portugal, Qatar, Republic of Korea, Republic of Moldova, Romania, Russian Federation, Rwanda, Saint Kitts and Nevis, Saint Lucia, Saint Vincent and the Grenadines, Samoa, San Marino, Sao Tome and Principe, Saudi Arabia, Senegal, Serbia, Seychelles, Sierra Leone, Singapore, Slovakia, Slovenia, South Africa, Spain, Sri Lanka, Sudan, Sweden, Switzerland, Syrian Arab Republic, Tajikistan, Thailand, Togo, Trinidad and Tobago, Tunisia, Turkey, Turkmenistan, Uganda, Ukraine, United Arab Emirates, United Kingdom, United Republic of Tanzania, United States of America, Uzbekistan, Viet Nam, Zambia, and Zimbabwe, June 19, 1970, modified October 3, 2001. As of December 7, 2021:
https://www.wipo.int/pct/en/

Pearson, Natalie Obiko, "Did a Chinese Hack Kill Canada's Greatest Tech Company?" *Bloomberg Businessweek*, July 1, 2020.

Pentheroudakis, Chryssoula, founder and managing director, IP Vanguard, "Technical and Practical Aspects Related to Patent Quality in the Context of Standard Essential Patents," commissioned by the World Intellectual Property Organization, undated. As of December 6, 2021:
https://www.wipo.int/edocs/mdocs/scp/en/wipo_is_ip_ge_18/wipo_is_ip_ge_18_a_study.pdf

Permanent Subcommittee on Investigations, Committee on Homeland Security and Governmental Affairs, U.S. Senate, *Threats to U.S. Networks: Oversight of Chinese Government-Owned Carriers*, staff report, Washington, D.C., June 9, 2020. As of December 6, 2021:
https://www.hsgac.senate.gov/subcommittees/investigations/hearings/majority-and-minority-staff-report_-threats-to-us-networks-oversight-of-chinese-government-owned-carriers

Public Law 115-232, John S. McCain National Defense Authorization Act for Fiscal Year 2019, August 13, 2018. As of December 7, 2021:
https://www.govinfo.gov/app/details/PLAW-115publ232

Qualcomm, "Everything You Need to Know About 5G," webpage, undated. As of December 5, 2021:
https://www.qualcomm.com/5g/what-is-5g

———, "Flagship Qualcomm Snapdragon 865 5G Mobile Platform Powers First Wave of 2020 5G Smartphones," press release, San Diego, Calif., February 25, 2020. As of June 11, 2020:
https://www.qualcomm.com/news/releases/2020/02/25/flagship-qualcomm-snapdragon-865-5g-mobile-platform-powers-first-wave-2020

Recon Analytics, "How America's 4G Leadership Propelled the U.S. Economy," Washington, D.C.: CTIA, April 16, 2018. As of June 11, 2020:
https://api.ctia.org/wp-content/uploads/2018/04/Recon-Analytics_How-Americas-4G-Leadership-Propelled-US-Economy_2018.pdf

Rubin, Alex, Alan Omar Loera Martinez, Jake Dow, and Anna Puglisi, "The Huawei Moment," CSET, July 2021. As of December 6, 2021:
https://cset.georgetown.edu/publication/the-huawei-moment/

Samsung, "Samsung Electronics Begins Mass Production at New EUV Manufacturing Line," press release, Korea, February 20, 2020. As of May 21, 2020:
https://news.samsung.com/global/samsung-electronics-begins-mass-production-at-new-euv-manufacturing-line

Semiconductor Manufacturing International Corporation, "About Us," webpage, undated. As of May 21, 2020:
https://www.smics.com/en/site/about_summary

Shaffer, Alan R., "A Microelectronic 'Canary in a Coal Mine': A Call to a New Approach for National Security," Potomac Institute for Policy Studies, April 12, 2021. As of December 6, 2021:
https://potomacinstitute.org/featured/2432-a-microelectronic-canary-in-a-coal-mine-a-call-to-a-new-approach-for-national-security

Shepardson, David, Karen Freifeld, and Alexandra Alper, "U.S. Moves to Cut Huawei Off from Global Chip Suppliers as China Eyes Retaliation," Reuters, May 16, 2020. As of June 11, 2020:
https://www.reuters.com/article/us-usa-huawei-tech-exclusive-idUSKBN22R1KC

Shilov, Anton, and Ian Cutress, "GlobalFoundries Stops All 7nm Development: Opts to Focus on Specialized Processes," *AnandTech*, August 27, 2018. As of May 21, 2020:
https://www.anandtech.com/show/13277/globalfoundries-stops-all-7nm-development

SMIC—*See* Semiconductor Manufacturing International Corporation.

Statcounter, "Mobile Operating System Market Share Worldwide," undated. As of May 22, 2020:
https://gs.statcounter.com/os-market-share/mobile/worldwide/

Statista, "Top Semiconductor Foundries Market Revenue Share from 2017 to the First Quarter of 2020," March 2020.

Strumpf, Dan, "ZTE's State Owner to Cut Its Stake," *Wall Street Journal*, March 13, 2019a.

———, "Huawei's Revenue Hits Record $122 Billion in 2019 Despite U.S. Campaign," *Wall Street Journal*, December 30, 2019b.

———, "China's Huawei Reports 38% Revenue Drop as U.S. Sanctions Bite," *Wall Street Journal*, August 6, 2021.

Taffet, Richard S., "Smartphone Patent Litigation and Standard Essential Patents," *Landslide*, Vol. 8, No. 4, March–April 2016, pp. 50–55.

Taiwan Semiconductor Manufacturing Company, homepage, undated a. As of May 21, 2020:
https://www.tsmc.com/english/default.htm

———, "5nm Technology," webpage, undated b. As of December 6, 2021:
https://www.tsmc.com/english/dedicatedFoundry/technology/logic/l_5nm

———, "7nm Technology," webpage, undated c. As of December 6, 2021:
https://www.tsmc.com/english/dedicatedFoundry/technology/logic/l_7nm

Takano, Atsushi, "Inside Huawei's First 5G Phone: Teardown Reveals Rush to Innovate," *Nikkei Asian Review*, October 23, 2019. As of June 1, 2020:
https://asia.nikkei.com/Spotlight/5G-networks/
Inside-Huawei-s-first-5G-phone-Teardown-reveals-rush-to-innovate

TelecomLead.com, "IHS Reveals LTE Infrastructure Market Share for Q1 2016," June 3, 2016. As of June 11, 2020:
https://www.telecomlead.com/telecom-statistics/ihs-reveals-lte-infrastructure-market-share-q1-2016-69184

Thompson, Loren, "Qualcomm Antitrust Case Raises Far-Reaching National Security Concerns," *Forbes*, January 28, 2020. As of January 31, 2020:
https://www.forbes.com/sites/lorenthompson/2020/01/28/
qualcomm-antitrust-case-raises-far-reaching-national-security-concerns/

Thurm, Scott, "Huawei Admits Copying Code from Cisco in Router Software," *Wall Street Journal*, March 24, 2003.

"Timeline: Chinese Telecoms Giants Huawei, ZTE Incur Wrath of Washington over Iran Sanction Violations," *South China Morning Post*, December 6, 2018.

Ting-Fang, Cheng, and Lauly Li, "Huawei Drops 5G for New P50 Phones as US Sanctions Grip," *Nikkei Asia*, updated July 29, 2021.

TSMC—*See* Taiwan Semiconductor Manufacturing Company.

UMC—*See* United Microelectronics Corporation.

United Microelectronics Corporation, homepage, undated. As of May 20, 2020:
http://www.umc.com/English/

United States v. Huawei Technologies, E.D.N.Y., superseding indictment, docket 18-CR-457 (AMD), filed February 13, 2020.

U.S. House of Representatives, CHIPS for America Act, H.R. 7178, 116th Congress, referred to the Committee on Science, Space, and Technology; the Committee on Ways and Means; the Committee on Armed Services; the Committee on Financial Services; the Committee on Energy and Commerce; and the Committee on Foreign Affairs, June 11, 2020. As of December 5, 2021:
https://www.congress.gov/bill/116th-congress/house-bill/7178/actions

U.S. Patent and Trademark Office, "About Us," webpage, undated. As of December 5, 2021:
https://www.uspto.gov/about-us

———, "Trademarks and Patents in China: The Impact of Non-Market Factors on Filing Trends and IP Systems," January 13, 2021a. As of December 6, 2021:
https://www.uspto.gov/about-us/news-updates/
breaking-news-uspto-report-examines-impact-chinese-government-subsidies-and

———, "Patents Data, at a Glance," published November 17, 2014, last updated June 8, 2021b. As of September 18, 2021:
https://www.uspto.gov/dashboard/patents/

U.S. Senate, CHIPS for America Act, S.3933, 116th Congress, referred to the Committee on Finance, June 10, 2020. As of December 5, 2021:
https://www.congress.gov/bill/116th-congress/senate-bill/3933

USPTO—*See* U.S. Patent and Trademark Office.

Vaswani, Karishma, "Huawei: The Story of a Controversial Company," BBC News, March 6, 2019. As of December 15, 2019:
https://www.bbc.co.uk/news/resources/idt-sh/Huawei

Verizon, "Verizon and AWS: A 5G Edge Cloud Computing Dream Team," webpage, undated. As of December 7, 2021:
https://www.verizon.com/business/solutions/5g/edge-computing/aws-verizon-edge-computing-announcement/

Virki, Tarmo, "Nokia Draws $560 Million R&D Loan for 5G," Reuters, March 6, 2020. As of December 5, 2021:
https://www.reuters.com/article/us-nokia-loan/nokia-draws-560-million-rd-loan-for-5g-idUSKBN20S339

Vodafone, "Vodafone Makes UK's First Holographic Call Using 5G," press release, September 20, 2018. As of January 19, 2022:
https://newscentre.vodafone.co.uk/press-release/vodafone-makes-uks-first-holographic-call-using-5g/

WG on Trust and Security in 5G Networks—*See* Working Group on Trust and Security in 5G Networks.

White House, *National Strategy to Secure 5G of the United States of America*, Washington, D.C., March 2020. As of December 5, 2021:
https://www.hsdl.org/?abstract&did=835776

Williams, Chris, "Having Swallowed Its Pride and Started Again with 10nm Chips, Intel Teases Features in These 2019-Ish Processors," *The Register*, December 12, 2018. As of May 21, 2020:
https://www.theregister.co.uk/2018/12/12/intel_architecture_future/

Woo, Stu, and Kate O'Keeffe, "Washington Asks Allies to Drop Huawei," *Wall Street Journal*, November 23, 2018.

Working Group on Trust and Security in 5G Networks, Center for Strategic and International Studies, *Accelerating 5G in the United States: Executive Summary*, Washington, D.C., March 1, 2021. As of December 6, 2021:
https://www.csis.org/analysis/accelerating-5g-united-states

Yap, Chuin-Wei, "State Support Helped Fuel Huawei's Global Rise," *Wall Street Journal*, December 25, 2019.

Zhu, Julie, "Exclusive: Huawei to Sell Phone Unit for $15 Billion to Shenzhen Government, Digital China, Others—Sources," Reuters, November 10, 2020. As of December 5, 2021:
https://www.reuters.com/article/huawei-m-a-digital-china-exclusive/exclusive-huawei-to-sell-phone-unit-for-15-billion-to-shenzhen-government-digital-china-others-sources-idUSKBN27Q0HJ

ZTE, "ZTE's Patented Technology Value Exceeds RMB 45 Billion," press release, Shenzhen, April 26, 2021.